Our Germans

Our Germans

Project Paperclip and the National Security State

BRIAN E. CRIM

Johns Hopkins University Press
Baltimore

Printed in the United States of America on acid-free paper

Johns Hopkins Paperback edition, 2020
9 8 7 6 5 4 3 2 1

Johns Hopkins University Press
2715 North Charles Street
Baltimore, Maryland 21218-4363
www.press.jhu.edu

The Library of Congress has cataloged the hardcover edition of this book as follows:
Names: Crim, Brian E., author.
Title: Our Germans : Project Paperclip and the national security state / Brian E. Crim.
Description: Baltimore : Johns Hopkins University Press, [2018] | Includes bibliographical references and index.
Identifiers: LCCN 2017016624| ISBN 9781421424392 (hardcover : alk. paper) | ISBN 9781421424408 (electronic) | ISBN 1421424398 (hardcover : alk. paper) | ISBN 1421424401 (electronic)
Subjects: LCSH: World War, 1939–1945—Technology. | Brain drain—Germany—History—20th century. | Scientists—Recruiting—Germany—History—20th century. | Scientists—Recruiting—United States—History—20th century. | Nazis—History—20th century. | War criminals—Germany—History—20th century. | Intelligence service—United States—History—20th century. | Military research—History—20th century. | German Americans—History—20th century.
Classification: LCC D810.S2 C784 2018 | DDC 940.54/867308850943—dc23
LC record available at https://lccn.loc.gov/2017016624

A catalog record for this book is available from the British Library.

ISBN-13: 978-1-4214-3818-4
ISBN-10: 1-4214-3818-6

Johns Hopkins University Press uses environmentally friendly book materials, including recycled text paper that is composed of at least 30 percent post-consumer waste, whenever possible.

CONTENTS

FIGURES

ACKNOWLEDGMENTS

This book began as a seminar paper in an elective diplomatic history course taught by Warren Kimball and Lloyd Gardner at Rutgers University during my first semester in the doctoral program. There is no doubt it was the worst paper I wrote in my graduate career, but my fascination with the subject endured. I am beholden to Warren Kimball in particular for savaging my work, deservedly so, and sustaining my interest in this most unusual episode in American history.

Omer Bartov continues to inspire my research. His thoughtful, reflective, and compassionate approach to some of the darkest periods in history is a model for my own work, and I am thankful to have studied with him. Carl Boyd is a great friend and mentor in addition to an excellent historian of intelligence during the Second World War. I continue to seek his advice on all things. I am incredibly grateful to the family of Walter Jessel for granting me access to his personal diary and memoir and for allowing me to share his story with the public.

Numerous institutions over the years provided support for this book, specifically Caldwell University, Lynchburg College's Faculty Development Committee, the Knight-Capron Library, especially Ariel Myers, and the John Mills Turner Distinguished Chair in the Humanities. The Virginia Foundation for Independent Colleges Mednick Memorial Fellowship Grant sustained a summer of research in College Park.

Asif Siddiqi, Seymour Goodman, Barton Hacker, Michael Neufeld, and Gerald Steinacher provided invaluable comments and research advice. Richard Koenigsberg and Orion Andersen invited me to share my work in the *Library of Social Science Newsletter*. The skilled professionals at the National Archives and Records Administration in College Park and the US Holocaust Memorial Museum simplified my life greatly.

It is a dream for any historian to have their work published by a press with the stature of Johns Hopkins. I am thankful Elizabeth Demers saw the potential in this work and appreciate her vote of confidence. Steve Robertson, Tom Fatouros, Dan Turner, and Lee Wright offered their friendship and often a place to stay while visiting the Washington, DC, area.

Most importantly, I am thankful for a loving and supportive family, without whom this book would not be possible.

ABBREVIATIONS

AAF	Army Air Force
ABMA	Army Ballistic Military Agency
ACA	Allied Control Authority
ACC	Allied Control Council
AMC	Air Material Command
BDC	Berlin Documentation Center
BIOS	British Intelligence Objectives Subcommittee
CIA	Central Intelligence Agency
CIC	Counter-Intelligence Corps
CIOS	Combined Intelligence Objectives Subcommittee
CROWCRASS	Central Registry of War Criminals and Security Suspects
DP	displaced person
EPES	Enemy Personnel Exploitation Section
EUCOM	European Command
FAS	Federation of American Scientists
FBI	Federal Bureau of Investigation
FEA	Foreign Economic Administration
FIAT	Field Intelligence Agency, Technical
FOIA	Freedom of Information Act
FRG	Federal Republic of Germany
HICOG	High Commission for Germany, Department of State
ICBM	intercontinental ballistic missile
INS	Immigration and Naturalization Service
IWG	Interagency Working Group
JCS	Joint Chiefs of Staff
JIC	Joint Intelligence Committee
JIOA	Joint Intelligence Objectives Agency

LUSTY	Luftwaffe secret technology
MIC	military-industrial complex
MVD	Ministry of the Interior, Soviet Union
NAACP	National Association for the Advancement of Colored People
NASA	National Aeronautics and Space Administration
NKVD	The People's Commissariat for Internal Affairs, Soviet Union
NSA	National Security Act (1947)
NSC	National Security Council
NSDAP	National Socialist German Workers' Party
NSF	National Science Foundation
NSRB	National Security Resources Board
OMGUS	Office of Military Government, United States
ONR	Office of Naval Research
OSI	Office of Special Investigations, Department of Justice
OSRD	Office of Scientific Research and Development
OSS	Office of Strategic Services
OTS	Office of Technical Services, Department of Commerce
POW	prisoners of war
RAF	Royal Air Force
RNC	Republican National Committee
SA	Sturmabteilung, "Storm Detachment"
SCAP	Supreme Command for the Allied Powers (Japan)
SHAEF	Supreme Headquarters Allied Expeditionary Force
SRD	Scientific Research Division, High Commission for Germany
SS	Schutzstaffel, "Protection Squad"
SVAG	Soviet Military Administration in Germany
SWNCC	State-War-Navy Coordinating Committee
V-1	*Vergeltungswaffe* 1, "retribution weapon"
V-2	*Vergeltungswaffe* 2, "retribution weapon"

Our Germans

Introduction

> The question discussed . . . is not whether we like or hate the Germans, scientists or plain citizens. Or whether we want or don't want them in this country. It's a question of what and how much these particular Germans can contribute to our scientific progress in a highly scientific age. In my opinion, we are entitled to exploit these talents to our best possible advantage.
>
> —Senator Harry F. Byrd

> Should Dr. Einstein[,] when he wrote down his famous formula for the relation between matter and energy, have dropped his pencil in despair because he had the vision of the release of unthinkably large amounts of atomic energy? Should Otto Lilienthal have discontinued his heroic glider flights because the possibility of a military misuse of the still unborn motorized airplane dawned on him? And should we rocket builders of today stop our efforts to open the universe to human exploration because rockets, just like airplanes, can be used also for military purposes?
>
> —Wernher von Braun

The story of Wernher von Braun and his rocket team's harrowing escape from the collapsing Third Reich into the warm embrace of a former enemy is one of the more exciting and significant moments from the final days of World War II. As the Red Army approached the Peenemünde research complex on the Baltic coast, the SS (Schutzstaffel), which had assumed control over the V-2 ballistic missile program in July 1944, evacuated valuable personnel and machinery into the interior of the country. Fearing the deteriorating regime would rather liquidate the rocket team than let them fall into enemy hands,

Wernher von Braun and other rocket team members surrendering to the US Army Counter-Intelligence Corps (CIC) in Ruette, Bavaria [Austrian Tyrol], on May 2, 1945. *Left to right,* Charles Stewart, CIC agent; Herbert Axster; Dieter Huzel; Wernher von Braun (in cast); Magnus von Braun (Wernher's brother); and Hans Lindenberg. *Source:* NASA/Marshall Space Flight Center

von Braun and his inner circle of specialists and loyal subordinates disappeared into the Bavarian countryside and patiently awaited contact with American troops.[1] It was the beginning of a long and fruitful relationship. More than the freighters full of equipment and caches of documents recovered from caves and hastily abandoned warehouses, the "German brains" who had designed and built the V-2 rocket along with dozens of other weapons in various stages of development proved invaluable to America's emerging military-industrial complex (MIC). The rocket team's contribution to the military during the Cold War and its subsequent role in the US space program are worthy of the enormous scholarship devoted both to the scientists and to the controversial program that brought them to American soil: Project Paperclip. For all the interest generated by journalists, popular culture, and seemingly constant revelations involving Nazis working for US intelligence or hiding in plain sight, a new history of Paperclip, emphasizing its highly contentious origins and legacy, seems timely.

Shortly after the Soviet Union shocked the world by launching Sputnik in October 1957, comedian Bob Hope urged the country to take the apparent defeat in stride. "It just proves one thing," Hope quipped. "Their German rocket scientists were better than our German rocket scientists."[2] Project Paperclip brought approximately fifteen hundred German and Austrian scientists, engineers, and technicians, along with thousands of dependents, to the United States for long-term "exploitation," as both military and civilian officials termed it in memos and correspondence, and eventually employment in the armed services, civilian agencies, and defense industries. Project Paperclip officially concluded in September 1947, but successor programs continued until 1962. The program derived its name from Army Ordnance officers' practice of attaching a paperclip to the folders of rocket experts they wished to employ.[3] The paperclip signaled to investigators that they should expedite the background check and, presumably, disregard normally prohibitive findings. Government memos defined Paperclip internally as a program for the "procurement of preeminent German or Austrian scientists and specialists for utilization in the interest of national security by the technical services of the Departments of the Army, Navy, and Air Force." An expanded program called National Interest broadened the scope of Paperclip by recruiting foreign scientists "for the military, and industrial concern, educational institution or other civilian agency." Project 63, initiated in 1950, was a smaller and short-lived variant of Paperclip designed primarily to "remove selected German or Austrian scientists and specialists from areas subject to immediate conquest in the event of hostilities and to utilize them in the national interest within the United States."[4] Most Paperclippers traveled to the United States between 1945 and 1947, and ninety percent of these Germans, who were contracted between 1945 and 1952, became citizens and remained in the country permanently.[5] Compared with the Soviets, "our Germans" may not have been as plentiful as "their Germans," but in the 1950s, the Soviet Union returned the German scientists they had employed, while scientists residing in the United States remained embedded in every layer of the MIC for decades.

In this book, I explore the process of integrating German scientists into the national security state emerging in the first decade of the Cold War. To do so, I address five interrelated topics. Using personnel files, from the rocket team primarily, I first delve into the backgrounds of scientists during the Third Reich and investigate the range of motivations driving Paperclippers' "second act" in the United States. German scientists prolonged the war through their efforts and prospered under the Nazi regime. Some were uncritical supporters of National Socialism, others more passive, but none of the Paperclippers achieved the impressive results so attractive to US authorities by opting out of the regime. Second, I describe the process by which the United States decided to obtain and integrate Nazi science and technology into the MIC. I also examine what US investigative authorities knew about their targets and, in many cases, how senior-level officials altered or concealed scientists' security dossiers to expedite immigration and circumvent the nation's laws.

The third topic concerns the vicious interservice and bureaucratic rivalries over resources and policy decisions relating to Paperclip. The most rancorous and sustained conflict pitted the Joint Intelligence Objectives Agency (JIOA), the agency responsible for Paperclip, against the State Department. What began as a necessary debate over immigration procedures devolved into a vindictive campaign to defame those who questioned Paperclip's rationale and to diminish the role of the department in national security policy. Senator Joseph McCarthy and his allies throughout the government, including FBI director J. Edgar Hoover, resurrected the Paperclip disputes of 1945–47 to discredit and slander State Department employees years later. Several interest groups, ranging from Jewish American organizations to the Federation of American Scientists (FAS), expressed outrage over rewarding Nazi scientists for their expertise in mass destruction. Although most Americans opposed Paperclip on principle, the initial negative publicity dissipated, and the program proceeded unfettered years after its public disclosure.

Fourth, several Paperclippers and the officers administering the program influenced the perception of the Soviet threat during the first decade of the Cold War. The frantic search for scientists consumed Allied intelligence resources and contributed to increasing hostility between the US and Soviet Union even before the German surrender. The Soviet Union claimed that scientists and materials in its zone constituted reparations, but so too did the French and British. Even if one questioned the benefits of Paperclip for the MIC or rejected it on ethical terms, denying the Soviet Union experts in military technology was an obvious priority for the national security state. Paperclippers returning from the Soviet zone fed US intelligence agencies alarming stories about Soviet advances in aeronautics, chemical and biological weapons, nuclear power, and guided missile technology. According to the reporting, Soviet military science benefited both from thousands of German prisoners and from a mutually beneficial arrangement between well-compensated German specialists and Russian talent. The Soviet threat appeared more menacing after debriefing German scientists, most of whom had a vested interest in promoting this perception to secure employment in the United States.

The final topic concerns the maturation of the national security state and the continuous involvement of Paperclippers in the MIC from the late 1940s through the 1950s. Hundreds of German scientists helped build, manage, and direct the MIC during this most crucial Cold War decade. I conclude by exploring the embattled legacy of Project Paperclip, a legacy that continues to arouse passions decades after the Cold War. How we choose to remember and represent Paperclip in culture and national memory, whether as an expedient Cold War program born from military necessity or a dishonorable episode, reflects our ambivalence about the MIC and the viability of an "ends justifies the means" solution to external threats. Wernher von Braun's rhetorical question comparing his own life's work to Albert Einstein's is both provocative and disingenuous, principally because von Braun was a willing and dangerously effective architect of one of the Third Reich's revolutionary weapons programs. The rocket team brought more than talent to the United States. Von Braun had perfected a managerial style that transitioned easily to the MIC. "The V-2," Walter McDougall notes, "was . . . a prototype of the national integration of brain power and material characteristics of the technocratic state."[6] Now provided with seemingly unlimited resources and the confidence of the state, the Paperclippers replicated their previous successes in the United States on a grander scale.[7]

Beginning in the mid-1980s, investigative journalists, such as Tom Bower, Linda Hunt, and most recently Annie Jacobsen, revealed that hundreds of Paperclippers actively supported the Nazi Party or its affiliates, including the SS, and a smaller number participated in some of the worst war crimes of the Holocaust.[8] Decades of reporting and scholarship divulged the willful subversion of immigration laws and denazification procedures by the US military and several civilian agencies in a coordinated effort to secure what the Joint Chiefs of Staff (JCS) deemed "chosen, rare minds" before foreign powers, principally the Soviet Union, did the same. Much of the literature on Paperclip is reductive and oversimplified. One can express continued moral outrage over the government's brazen duplicity and deception by providing a haven for the most unsavory characters, or conversely, venerate and chronicle German science's exceptional contribution to America's Cold War victories and storied space program. Neither approach is particularly enlightening.

Nuanced scholarship about the exploitation of German science abounds, and the historiography regarding Nazi science and technology is certainly rich, but much more remains to be written on the topic of Paperclip thanks to the release of millions of documents by the Nazi War Crimes and Japanese Imperial Government Records Interagency Working Group (IWG) in 2007.[9] The question "why did the United States hire ex-Nazis?" is answered, even if the response is exasperating. How hundreds of ex-Nazi scientists contributed to the national security state, whether as technocrats or intelligence assets, deserves greater scrutiny.

Senator Harry Byrd's 1948 article defending Paperclip exemplified the prevailing logic within government and military circles—whether we like or even trust these Germans completely is immaterial. What can they do for us? Moreover, how can we risk abandoning this pool of scientific and technical talent to a less discriminating enemy, one poised to overrun Western Europe at a moment's notice? These considerations won the day, but they obscure the fact that Paperclip was a contentious program within the US bureaucracy. It is easy in retrospect to delineate the triumph of expedient political calculations in virtually every Cold War episode and policy formulation, but forging a consensus on national security questions in the early post–World War II period involved intense power struggles and sometime hostile takeovers. Project Paperclip and its successor programs coincided with the collapse of the Third Reich and the start of the Cold War. As such, it is an excellent microcosm with which to explore the conflicting worldviews present during the formation of the national security state. Furthermore, hundreds of Paperclip recruits integrated seamlessly into the MIC, some within just months of arriving in the United States. Whether they remained under military employment, transitioned to civilian agencies like the National Aeronautics and Space Administration (NASA), or sought more lucrative careers with corporations flush with government contracts, German specialists exercised significant influence over the national security state.

Science and "National Security"

Few in the US questioned the need for a permanent and global military presence after the cessation of hostilities in Europe and the Pacific. By autumn 1945, prominent civilian and military officials began articulating a postwar agenda in which they envisioned a closer link between military power and foreign policy. Military planners advocated integrating the military, the economy, and science and technology in the furtherance of what they termed "national security."[10] A War Department official defined the requirements for national security as an enhanced intelligence capability, a "national realism that will permit us to start all our preparations for war when we see aggressive intent in another nation," as well as an improved ability to mobilize national resources. Navy secretary and future secretary of defense James Forrestal echoed this logic, telling Congress, "I am using the word 'security' here consistently and continuously rather than 'defense.'"[11] Daniel Yergin notes that the doctrine of national security rested on four precepts. First, the Eurocentric international system had collapsed and been replaced by a bipolar system dominated by the United States and the Soviet Union. Second, the experience of global total warfare was not an aberration but the future of conflict. Third, the United States required a powerful and balanced military machine to ensure credibility abroad. Fourth, technology drives military preparedness, and America's penchant for isolationism, in the words of Admiral Randall Jacobs, "had been repealed by science."[12] "Imagine a war that began with Polish cavalry charging German tanks," writes Yergin, "ended with the V-2 over London and the atomic bomb." The apparatus that delivered victory in World War II must remain in place, national security proponents claimed. National security doctrine assumed a permanent enemy capable of threatening the United States. Whether the Soviet Union posed a legitimate threat was irrelevant; it gave national security doctrine "purpose and urgency."[13] US intelligence services struggled to evaluate Soviet military capabilities accurately during the early postwar era. Inflated estimates produced by the intelligence community suggests that the United States not only overvalued Soviet abilities, but perceived itself vulnerable and powerless over world events.[14]

The institutional structure of national security originated with the need to coordinate with Britain during World War II and provide the president a responsive politico-military bureaucracy. The JCS, which duplicated the British staff model, and the State-War-Navy Coordinating Committee (SWNCC) ensured that military planning and postwar planning proceeded in unison, at least on paper. The challenges of occupation and legitimate concerns over Soviet intentions in central Europe precluded dismantling these structures and, after a short time, required a more ambitious reorganization. The National Security Act (NSA) of 1947, Thomas Etzold writes, responded to three lessons learned from the Second World War: "coordination of political and military policy; improved intelligence capability; and rationalization of the defense community."[15] While organizations like the National Security Council (NSC) proved its worth over decades, other institutions created by the NSA responsible for rationalization failed to prevent duplication of effort, competition over strategic missions, and bitter interservice rivalries over resources. Paperclip fell under the purview of the JCS via the JIOA and evolved into  something more than a procurement effort once the scientists dissolved into the sprawling national security bureaucracy. Each armed service and the growing number of defense industries sought these prized Germans and afforded them significant power and resources.

The doctrine of national security, or what Michael Sherry termed an "ideology of national preparedness," signaled the militarization of foreign policy and an era of expanding budgets.[16] The architecture of the national security state, portions of which were erected during the Second World War, directed spending, apportioned administrative control, and exercised inordinate influence over economic activity.[17] Disagreements between the military and civilian authorities over occupation policy and interpreting Soviet behavior  and capabilities in the immediate postwar years bloomed into a larger conflict over defining national interest and long-term security objectives. National security privileged military planners over traditional civilian foreign policy makers. Pentagon officials pursued their own objectives at home and abroad with little regard for their counterparts at the State Department.[18] Furthermore, several flag officers assumed upper echelon positions at the State Department during the early postwar years, George C. Marshall most notably. This practice promoted national security doctrine on the macro and micro levels of policy formulation throughout the bureaucracy. Thomas Etzold notes that the period 1945–50 "were years of drastic decline for the State Department as an institution." Both secretaries of state Marshall and Dean Acheson played a  role in ceding responsibility for Paperclip's immigration question to the military with predictable results. Marshall and Acheson, Etzold writes, "proved influential not only with the President but within the NSC system."[19] The State Department ultimately transitioned into an arm of the national security state.

Given the declining status of the State Department in the early Cold War period, it is ironic that the incredibly influential NSC-68 report, entitled *United States Objectives and Programs for National Security*, originated with the State Department's policy planning staff. Drafted months after the Communist victory in China and the testing of the Soviet atomic bomb, the April 1950 study called for "the rapid building up of the political, economic, and military strength of the free world." Paul Nitze, the principal author, depicted the Cold War as a Manichean conflict between freedom versus slavery, in which the Kremlin sought nothing less than "the complete subversion or forcible destruction" of Western institutions and the "domination of the Eurasian landmass" through its enormous military resources and growing nuclear arsenal. The US response had to match Soviet power with its own robust permanent military institutions. US strength, the report noted, depended on "an accelerated exploitation of [American] scientific potential."[20] Implementing NSC-68 doubled the size of the armed forces and increased defense appropriations by 262 percent.[21] For Paperclippers residing in the United States in 1950, the sky was the limit.

The centrality of advanced research and development to national security policy elevated scientists to new heights of power and influence. The foundation for scientists' inclusion in postwar policy circles derived from such triumphs as the Manhattan Project, certainly, but several scientists adroitly positioned themselves as spokesmen for "national preparedness" in the postwar period. "Scientists," Michael Sherry writes, "did not drift aimlessly into military research, nor were they duped by it. They espoused its virtues, lobbied hard for it, and rarely questioned it."[22] With thousands of Germans enlisted in Soviet military research, voluntary or otherwise, national security apostles in the United States had little trouble sounding the alarm and depicting apocalyptic scenarios. Vannevar Bush, the engineering and administrative genius behind the Office of Scientific and Research Development (OSRD) and author of the influential 1945 report *Science: The Endless Frontier*, endorsed substantial national security infrastructure mirroring the OSRD, which historians rightly judge as "the most centralized and highly integrated administrative agency for the mobilization of science and technology to have emerged during the Second World War."[23] The man who organized the Manhattan Project, advised presidents, and pioneered "Big Science" commanded an audience. Bush invoked recent history and the menacing Soviet presence in Europe to justify a permanent version of the OSRD in peacetime—the National Science Foundation (NSF): "We must grasp the tough fact that the very emphasis on peace in the great democracies, in the interval between the last two wars, undoubtedly fostered the aggressors' conviction that the democracies were soft and decadent, and encouraged Hitler to strike. Talk of peace must this time be realistic; we shall need to maintain our full strength as a military power if we are to be respected and listened to."[24]

In the era of the atomic bomb and the ballistic missile, formulating strategy demanded scientific input to keep abreast of technological advances. Bush modeled the NSF in part on the Office of Naval Research (ONR), which encouraged "scientific research in recognition of the paramount importance [in] the preservation of national security." The NSF's mission statement declared science had to both "advance the national health, prosperity, and welfare" and "secure the national defense."[25] Vannevar Bush proposed this marriage between science and national security, and Wernher von Braun, Paperclip's most revered acquisition, embodied the scientist-advocate representative of the new national security environment.

Sociologist C. Wright Mills's enduring study of what he termed "the power elite" provides insight into science's role within the national security state and the MIC. Writing in 1956, Mills identified the "warlords" (military officers), the corporate leaders, and the political artifice, especially the bureaucracies, as elite men who pulled the various levers of power. "By the power elite," Mills writes, "we refer to those political, economic, and military circles which as an intricate set of overlapping cliques share decisions having at least national consequences. In so far as national events are decided, the power elite are those who decide them."[26] Mills emphasized that military funding was responsible for the majority of "pure scientific research" and most basic research as well. As of 1954, Mills calculated, eighty-five percent of government scientific research fell under the umbrella of "national security."[27] Scientists relied on this "preparedness" spending and clearly benefited from the expanding scientific infrastructure, along with closer ties between defense industries and policy makers. The political administration in power was irrelevant—national security, like the containment doctrine underlying it, took on a life of its own. Despite Paperclippers' collaboration with the Third Reich and complicity in its crimes, many joined the ranks of the power elite and flourished.[28]

The Military-Industrial Complex

James Ledbetter's definition of the MIC as "a network of public and private forces that combine a profit motive with the planning and implementation of strategic policy" is reflective of a reality deeply entrenched in the United States at the time of Dwight Eisenhower's Farewell Address in January 1961.[29] Eisenhower warned the public about the "unwarranted influence" of the MIC and

seemed to channel Mills's observations about a shadowy power elite controlling the agenda: "[I]n holding scientific research and discovery in respect, as we should, we must also be alert to the equal and opposite danger that public policy itself become the captive of a scientific-technological elite."[30] The irony is rich given the administration's deliberate and thorough integration of science into the maturing national security infrastructure. Furthermore, Eisenhower deliberately surrounded himself with his own "scientific-technological elite" to outmaneuver the corporate MIC and guarantee that science served the national interest, not the profit motive.[31] James Killian, president of the Massachusetts Institute of Technology, continued the work of his predecessor, Vannevar Bush, by creating the Presidential Science Advisory Committee and became the first true science adviser to an American president. Killian's attitude toward Cold War science reflected a true statist approach. He wanted to "muster the democratic ranks of American scientists into invisible human battalions" and mobilize research universities into "a powerful fleet-in-being," capable of being "thrown instantly into action."[32] Such comments underscored the unquestioned premise that scientific advancements dictated strategy. Eisenhower did not so much disparage the existence of the MIC, which included agencies like the Atomic Energy Commission, Defense Department, and NSF, as lament the ascendance of corporate elites influencing government spending and national security policy.[33]

The sociological concept of the "iron triangle," describing the interrelationship between Congress, the bureaucracy, and interest groups, has an equivalent in the MIC. The "golden triangle," comprising military services and agencies, defense industries, and research universities, fostered the militarization of the economy and of broader aspects of American life.[34] The result, Big Science, served the national security state by focusing on behemoth projects, employing large numbers of scientists and normalizing massive expenditures. Big Science furthered national security and national prestige simultaneously.[35] Sociologist Harold Lasswell anticipated the MIC in 1941 by predicting that technology would outpace civilian societies' ability to control it, creating a permanent imbalance between military and civilian institutions. The "garrison state" might describe Nazi Germany and the Soviet Union, but Lasswell, like Mills, witnessed a similar trend in the democracies organized for total war. Eisenhower obviously read and agreed with Lasswell. "We don't want to become a garrison state," Eisenhower said. "We want to remain free. Our plans and programs have to conform to a free people, which means essentially a free economy."[36] Edward R. Barrett, the assistant secretary of state for public affairs in the Eisenhower administration, also cited Lasswell, arguing that "we are going to run into vast opposition among informed people to a huge arms race. We will be warned that we are heading towards a garrison state."[37] Lasswell also envisaged military professionals in the garrison state assuming

responsibilities "we have traditionally accepted as part of modern civilian management."[38] Both Mills's "power elite" and Lasswell's "specialists of violence" bear a striking resemblance to the Paperclippers who assimilated into the MIC, an observation that speaks more to continuities between the Third Reich's MIC and the US equivalent after World War II. Big Science knows no ideology.

Admiral Luis de Florez, an influential figure in the navy's Bureau of Aeronautics and the CIA's first director of technical research, bitterly concluded after World War II, "If only we had been smart enough to carry off the thousand best German scientists and technicians and shut them away on a scientific St. Helena, Europe would have remained disarmed for a generation."[39] While this logic seemed attractive in light of Germany's clandestine rearmament after World War I, by 1945 Allied military officers harbored decidedly different opinions. Fears of German resurgence remained, but even Paperclip's skeptics recognized German science had much to offer if properly vetted and employed by responsible actors. Michael Neufeld, noted biographer of Wernher von Braun and historian of the German rocket program, argues persuasively that Peenemünde's legacy was truly revolutionary. The rocket program "was one of the first examples of state mobilization of massive engineering and scientific resources for the forced invention of a radical, new military technology."[40] Aside from a reliance on slave labor, which is no small distinction, Neufeld finds more similarities between Peenemünde and Los Alamos than differences, including "ultra-secrecy, massive state investment, the need to scale up exotic technologies to an industrial level, and the harnessing of university research for weapons development."[41] The MIC of the Cold War period owed much to previous German accomplishments, and with the US acquisition of 1,500 scientists and technicians in the decade after World War II, Germans put their indelible stamp on America's greatest scientific endeavors of the second half of the twentieth century. The final report issued by the Field Intelligence Agency, Technical (FIAT) in 1947 estimated that the War and Navy Department acquisition of rockets, chemicals, and wind tunnels could "be measured in the billions of dollars," and the report predicted, "[T]he US government and industry will financially receive 1,000 times more value than it expended on the project."[42] Scientists recognized their inherent value and influence over strategy, economic activity, and public policy in no small part because of German contributions to military research and development on two continents.[43]

Paperclip's *Historikerstreit*

The West German *Historikerstreit* (historians' dispute) of the late 1980s describes an intense debate between right-wing intellectuals and left-wing intellectuals over the proper legacy of Nazi war crimes in contemporary Germany. The Right considered the Third Reich and Stalin's Soviet Union comparable since they were both totalitarian regimes, while the Left regarded National Socialism unique in its criminality.[44] The debate coincided with the collapse of the Soviet Union, euphoria over German unification, and renewed historical attention to Soviet offenses in Eastern Europe. Project Paperclip has its own embattled historiography in which those ideologically invested in its success and in preserving a positive legacy vied with investigative journalists, Holocaust survivors seeking a full accounting of Paperclippers' activities during the Third Reich, and historians mining new or declassified archival sources. Jean Michel, a survivor of the Dora concentration camp, which was used to supply labor for the underground V-2 production facility at Nordhausen, condemned "the monstrous distortion of history, which, in silencing certain facts and glorifying others, has given birth to false, foul and suspect myths."[45] Michel was outraged that "[m]en who were intimately involved with the creation and operation of the camp are today, venerated and admired."[46] The V-2 was not the "Immaculate Conception" of the Third Reich, Michel reminds us, but a terror weapon built on the corpses of tens of thousands of slave laborers.[47] These truths would take decades to come to light.

For nearly a half century, histories of the German missile program and Project Paperclip originated almost entirely from participants or supporters writing in the shadow of the Cold War.[48] Consequently, authors minimized or ignored altogether associations between the scientists and the Third Reich, even when some prominent Paperclippers were implicated directly in the use of slave labor. Separating myth from reality proved exceedingly difficult without the trove of documents unearthed by journalists or released by the IWG.[49] Wernher von Braun and Walter Dornberger, the Wehrmacht officer in charge of the V-2, were particularly adept at constructing a narrative exonerating their work on behalf of the Nazi regime, a narrative that served both the scientists' and the US military's interests.[50] As I note in chapter 1, von Braun carefully guarded information about himself, his associates, and their knowledge to garner the best possible concessions from the US Army. Once integrated into the MIC, the Army Ballistic Missile Agency (ABMA) and NASA public relations officials worked closely with the rocket team to manage their stories and to promote what Rip Bulkeley aptly named the Huntsville School of American space history. This school expresses the scientists' perspective, particularly von Braun's, that "ill-treated but dogged pioneer who eventually triumphed against the odds." The Huntsville School conveniently ignores the team's wartime activities and willfully neglects certain "social and political aspects" of the space program and its military antecedents.[51] Roger Launius, a NASA historian and curator at the National Air and Space Museum, decries the Huntsville School as "unabashedly celebratory" of the German contributions while minimizing significant American contributions like the Jet Propulsion Lab and other advances in rocketry.[52] Regardless of Cold War exigencies or reverence for NASA's accomplishments, specifically, the careers of some well-placed Paperclippers, Michael Neufeld reminds the scholarly community that exhaustively researching the backgrounds of the Germans "raises the most fundamental questions regarding the moral responsibility of scientists and engineers in the twentieth century—and twenty-first."[53] The purpose of problematizing the Huntsville School in this book is not to undermine the accomplishments of the space program, but to place Paperclip and other US policies in a post–Cold War context.[54]

Historians owe a debt to journalists Linda Hunt and Tom Bower for pursuing countless Freedom of Information Act (FOIA) requests and interviewing significant figures from the Paperclip program. Hunt donated her interviews and research materials to the US Holocaust Memorial Museum Archives, although much of what she collected in the 1980s is now declassified and archived in College Park, Maryland. As valuable as this journalistic work is, historians John Gimbel and Michael Neufeld are properly skeptical of accusations of a Paperclip "conspiracy" or "secret agenda." "They [Hunt and Bower] bash the Pentagon," Gimbel writes, "but give little or no attention to the broader base of the policy. That results now only in a distorted view of the policy itself, but also a false interpretation of the role of President Truman and several other agencies of the government." Paperclip was anything but a conspiratorial enterprise hatched by Machiavellian military officers bent on subverting laws, morality, or civilian control over the military. It was, Gimbel continues, "a national policy . . . developed and implemented by duly authorized, responsible agents of the US government."[55] Indeed, one of the more striking aspects of Paperclip is the widely shared assumption among policy makers that it was a wise and justified program. The JIOA thought nothing of blatantly violating existing immigration laws, mostly because its military directors assumed its mission trumped all other concerns. Gimbel is hardly dismissive of the troubling aspects of Paperclip; his focus is the genesis and implementation of American "exploitation," or plunder of German science during the occupation period.[56] Ruminating over the morality of Paperclip is a worthy exercise, as is the relentless pursuit of the truth behind Paperclippers' backgrounds, but the internecine warfare inside the bureaucracy and the complicated legacies of Paperclip require greater scholarly attention, especially in light of new sources.

The vicious war of words between the military and the State Department during Paperclip's formative years influenced the historiography of the project from the outset. Clarence Lasby began writing the first academic history of Paperclip in the 1960s, when most of the documents were classified, but many of the scientists were available for interviews. Lasby acknowledged working closely with the military to gain access, so much so that it must be said he promotes a pro-Paperclip perspective, mirroring the army's own official histories of the program. Moreover, Lasby attacked the State Department for being "truly obstructive." "They would not grant access to their documents," he writes in the preface, "because they fell within their 'closed period'—1944 and after; and they practiced the only real censorship, insisting to the end that I omit the name of their major policymaker in the program."[57] The "one man" Lasby refers to frequently in the book as Paperclip's nemesis is a key figure in my own—Samuel Klaus, a State Department lawyer assigned to the JIOA who questioned the legality and wisdom of Paperclip from the outset. The military and FBI granted Lasby access to material in part to continue its crusade against Klaus and to promote a positive history of their involvement in a controversial program. One 1961 FBI memo claimed that helping Lasby, a graduate student at the time, "could further strengthen our position in academic circles."[58] J. Edgar Hoover, once a fierce critic of Paperclip before succumbing to the national security argument presented to him, granted Lasby documents and access to scientists.[59] Lasby thanked Hoover with a note and a copy of the book, which predictably portrayed the FBI director as a pragmatic patriot. The State Department, at least the middle-level officials like Klaus, were silenced during Paperclip's formative phase, but they are also victims of a sterile and triumphant historiography vindicating Project Paperclip and denigrating its critics for shortsightedness.

"The tragedy of American diplomacy is not that it is evil," William Appleman Williams wrote in 1972, "but that it denies and subverts American ideas and ideals."[60] Paperclip was not evil, but it challenges narratives emphasizing American exceptionalism and our professed idealism. The program implemented in the shadow of the Cold War was more than an expedient solution; it belied a worldview privileging technocracy and a permanent national security state. Restoring balance to both the official record and the often politicized historiography of Paperclip motivated me to write this book. Paperclip encompasses fundamental questions about science and morality, but also reveals much about accountability, or lack thereof, even after the Cold War. The Department of Justice's Office of Special Investigations (OSI), which investigated several Paperclippers and instigated denaturalization procedures against former V-2 engineer and NASA executive Arthur Rudolph in 1982, published a report in 2006 as part of a transparent effort to answer troubling questions about the Cold War era. "Did we knowingly permit major or even minor Nazi persecutors to enter, and if so, what justification was given? At what level within the government was there legal and moral authority to advance such a policy? And were efforts made to conceal such activities from the public to advance some perceived higher national good?"[61] "Yes" is the answer to all, but the how and why remain elusive. Project Paperclip speaks to enduring issues, from the relationship between science and the state to the complexity of civil-military relations in an unsettled age.

CHAPTER ONE

Aristocracy of Evil

The Paperclippers and Nazi Science

> They were enthusiastic technicians with the mission according to Goebbels of saving Germany. As a team they were granted all the financial support, materials, and personnel they required, within the means of the German war machine. Continuance of the work depended on continued conduct of the war. At a time when the generals were dissatisfied with the party rule to the extent of attempting to overthrow it, Peenemünde was out of touch and sympathy with such developments—not for love of the party necessarily but because their work and the war were one.
>
> —Walter Jessel, Second Lieutenant, June 12, 1945

Walter Jessel, a German Jew who escaped Nazi Germany only to return as a US soldier assigned to Patton's Third Army and eventually to the Office of Strategic Services (OSS), had the distinction of interrogating German scientists and military officers during their first month of captivity.[1] Most of Jessel's subjects belonged to the Peenemünde rocket team, led by Wernher von Braun and General Walter Dornberger. Jessel was remarkably forthright in his assessment of these men, who seemed to intuit their enviable position as "prisoners" of an American military anxious to tap into their expertise. Rather than sing their praises, Jessel questioned the Germans' motives, their sincerity, even their knowledge, and concluded that their ideological fervor, or more accurately, amoral opportunism, would not dissipate just because the privileged Germans served new masters. If Jessel was unduly harsh in his evaluation, which is debatable, most American officers and civilians pushing for Paperclip were willfully ignorant of the attitudes and intentions of the hundreds of scientists to whom they offered employment and citizenship.

In this chapter, I first explore the "coordination" of scientists and engineers during the Third Reich, specifically the relationship between prominent scientists and the Nazi MIC. In this first section, I analyze the dynamics of the V-2 program and its relevance as an organizational model beyond the Third Reich. Second, I address scientists' complicity in Nazi atrocities or criminal programs and policies during World War II. Third, I construct a collective profile of the Paperclip scientists and engineers using several case studies and collected data from German and American sources. Most of the subjects are among the approximately 120 rocket scientists assigned originally to Peenemünde and the Nordhausen complex before making the journey to the United States. Building the profile requires understanding how the United States perceived Nazi science and scientists with an eye toward exploiting them for military and civilian use. Military and civilian agencies investigated targeted scientists and often differed on both the scope of the inquiry and the significance of incriminating information, but most agreed that a specialist's potential value to national security superseded all other concerns.

Rocket Scientists under the Swastika

German science was the envy of the world before World War I. Competing nations sent students to German universities to learn from a veritable army of Nobel Prize laureates in hopes of attaining higher standards of research and scientific education. The international scientific community understandably shattered after war devastated Europe, leaving German scientists isolated and monitored, along with the entirety of the country's war potential. German scientists were forbidden to participate in international meetings until 1922. Despite external restraints and profound economic limitations during the interwar period, most industrial sectors developed ancillary research and development institutes, which organized scientists and engineers into interdisciplinary teams.[2] This infrastructure facilitated the flurry of military research initiatives during the 1930s. Nationalist sentiment infected the scientific community no less than it did the broader German university population, and many prominent scientists gravitated to conservative and nationalist causes, including the Nazi Party.[3] Technical universities in particular, writes Michael Petersen, were "hothouses of a virulent strain of right-wing conservatism that stressed the ability of technology to lead the way to a renewal of the German spirit and nation."[4] Scientists contributed to the interwar rearmament effort orchestrated by the chief of staff of the Reichswehr Hans von Seeckt, whose covert efforts allowed military science to survive despite the restrictions leveled by the Treaty of Versailles (1919).[5] Starved for funding in a feeble German economy, university professors aligned their research agendas with armaments in order to attract financial backing. Young students, like Wernher von Braun and others who would join the rocket team, learned early on that the state was a reliable benefactor if their research included a military dimension.[6] The Third Reich did not provide the blank check scientists anticipated, but the possibilities for patronage and funding enthused a generation whose ambitions would otherwise be thwarted in the Weimar system. Historian Michael Wildt's study of the "generation of the unbound" argues that young, highly educated, and nationalist professionals like those in certain SS departments prospered under the Third Reich. Talented scientists with comparable backgrounds achieved similar success.[7]

Adolf Hitler articulated an ambivalent attitude toward the hard sciences even as he came to rely on them for both military and civilian pursuits. In *Mein Kampf* Hitler acknowledged the need for chemistry, physics, and mathematics in a technological era, but he cautioned against the "materialistic egoism" inherent in the sciences. *Bildung*, which Hitler considered a comprehensive moral education stressing sacrifice to the community, must be front and center in any profession.[8] The party's attitude toward science was utilitarian—"what was useful was good."[9] The Nazi Party began a systematic campaign of "coordination" (*Gleichschaltung*) of all national institutions and professions within months of taking power in 1933. Coordination demanded not only aligning one's values with the party, but also placing infrastructure and resources at the disposal of the state. Doing so often came with rewards in the form of funding and rapid promotion. Jews were the first casualties of coordination, usually at the hands of the profession being coordinated, not the state. Anticipating the party's racial policies, many professions purged Jews from membership rolls and stripped Jews of their livelihood unprompted. By 1935, approximately twenty percent of scientists were removed from their positions, more in fields like physics.[10] Non-Jews could also be "coordinated" out of a position if they failed to adapt to the new ideological reality.

Most historians agree that the German scientific community practiced a form of self-coordination to safeguard their research (and funding) and to retain some autonomy, although scientists were hardly immune to ideological concerns. Alan Beyerchen believed physicists "desired strongly to remain aloof from political concerns . . . but they were not able to do so."[11] Likewise, Margit Szöllösi-Janze claimed German scientists were "predominantly not ideologically dogmatic, but largely rational and results-oriented."[12] The very nature of the Third Reich encouraged hyperspecialization and rewarded results, but it is difficult to dispute Michael Petersen's conclusions regarding the rocket team in particular: "[T]hey came to see the concerns of outside groups as being far less consequential than their own. The result was a narrowed technical and patriotic vision that consented to some of the worst crimes of the Nazi regime."[13] Scientists integrated into the military-industrial-university complex embraced a version of technocratic amorality comparable to Wehrmacht officers who ceded responsibility for crafting strategy to a Nazi elite indifferent to rational means and ends.[14] These technocrats differed from their counterparts abroad for their willful assistance in furthering the untechnocratic goals of a reckless state.[15] In fact, the very irrationality of the state's military planning enabled dynamic technological revolutions like the V-2. Military officers and scientists did more than consent to the Third Reich, however—they counted among the perpetrators of its worst crimes.

Scientific mobilization during the Third Reich mirrored the functionalist model of interpreting the regime. According to this school of scholarship, Nazi Germany was characterized by multiple power centers revolving around key personalities and interests, which competed for influence over policy formation and resources. Hitler stood atop this polycratic empire of warring bureaucratic fiefdoms and encouraged competition by communicating his desires to rivals and choosing among options presented to him. This process of "cumulative radicalization" encouraged the opposing agencies and personalities to promise grandiose solutions to complex problems, thereby winning Hitler's approval.[16] Navigating the chaotic structure was a prerequisite for accomplishing anything of importance in the regime, whether it was the Final Solution, economic mobilization, or constructing a revolutionary weapons system. In every case, inefficiency reigned supreme, although the polycentric structure was not without significant breakthroughs. In 1939 the British determined that the Germans had made impressive gains in electronic guidance for rockets, torpedoes, and glider bombs. The Oslo Report, produced by a British official assigned to the embassy in Norway, provided early intelligence to the Allies on Peenemünde and the development of guided missiles.[17]

The Nazi regime initially drafted thousands of scientists and technical specialists indiscriminately, without first determining their value as military assets. This practice continued until June 1942, when Hitler issued a decree requiring "the concentrated effort of scientific research and its channelization toward the goal to be aspired." The decree proclaimed, "Leading men of science above all are to make research fruitful for warfare by working together in their special fields."[18] By the latter years of the war, the decentralized administrative system of weapons development gave way to Albert Speer's impressive efforts to consolidate armaments production. In June 1944, Hitler issued another decree privileging "weapons and equipment which by revolutionary new characteristics, will give us a sensible superiority over developments made by the enemy."[19] The V-2 was an irrational weapon, especially given the German military's shortages in more vital areas, but it benefited from Speer's rationalized production process and Hitler's demand for ever more fantastic "wonder weapons."[20] Predictably, the SS managed to carve out its own armaments empire and soon coveted the resources, brainpower, and potential associated with the Peenemünde complex.

The genesis and evolution of the V-2 program reveals much about the backgrounds of key Paperclip acquisitions, how portions of the program were replicated in different locations in the United States, and why the Peenemünde model of organization was incorporated in the postwar MIC. The relationship between Walter Dornberger, an officer in the German military's ordnance department who understood the potential in rocketry, and the young doctoral candidate Wernher von Braun helped translate an ambitious vision into a mass-produced weapon of war. Moreover, the two shepherded the V-2 program, primarily the brainpower behind it, to safety in the United States, where the relationship continued to produce impressive results for their new patrons. Dornberger first met von Braun in 1932 at the Society for Spaceship Travel (Verein für Raumschiffahrt) and offered him a chance to pursue rocketry and complete his degree at the University of Berlin. The twenty-year-old engineer became the civilian technical director of the army's embryonic rocket program.[21] The army ordnance rocket group built its own laboratory and spearheaded the Nazi "coordination" campaign in concert with the Gestapo and intelligence services, by either silencing independent rocket enthusiasts or co-opting them in the military's program.

General Dornberger aligned his group with Hermann Goering's Luftwaffe in 1935, and Peenemünde was up and running by early 1937.[22] Dornberger pressed for an "unter einem Dach" (under one roof) capability so that a project as complicated as the V-2 could move from the drawing board to the assembly plant with minimal interruption or outside interference.[23] Arthur Rudolph recalled the value of concentrating research in the government, not in private enterprise. "You didn't have all the administrative work in dealing with the contractor and the frictions along with it," Rudolph recalled, "That was all eliminated. So it became very effective. That was the goal of Dornberger."[24]

The team of Dornberger and von Braun forged alliances with civilian scientists and engineers in industry and universities, as well as military technocrats across the armed services.[25] There was no shortage of slave labor either. Dornberger was nothing if not "a talented salesman and bureaucratic in-fighter,"[26] as Neufeld describes him, but the real testament to the success of Peenemünde was best summarized by one of Dornberger's interrogators after the war, Caltech astrophysicist Fritz Zwicky, who described the V-2 as "a technical achievement of high order due less to the activity of any individual genius than to the determined and enthusiastic cooperation of a large number of only moderately competent technical individuals."[27] In other words, Zwicky concluded, with the right organizational mindset and outlay of resources, Peenemünde could be replicated anywhere.

Wernher von Braun's life and career is well documented, but it is worth exploring his vision for Peenemünde and assessing his administrative talents since both were essential for his postwar accomplishments in the United States. Von Braun's recollection of his first meeting with Dornberger is incredibly revealing. Surrounded by space enthusiasts searching for a home, von Braun welcomed Dornberger's offer of military patronage: "We felt no moral scruples about the possible future use of our brainchild. We were interested solely in exploring space. It was simply a question with us of how the golden cow could be milked most successfully."[28] Within a year of the meeting, the golden cow changed ownership, and the newly established Wehrmacht seemed to promise unlimited possibilities—and challenges. Dornberger may have maneuvered the V-2 program through the byzantine priority and procurement system and provided protection for his scientists from rival agencies like the SS for as long as possible, but von Braun managed the army of engineers assigned to the project.[29] Von Braun was, first and foremost, an excellent talent scout, traveling across Germany and inviting young engineers like himself to participate in a unique experiment.[30] V-2 engineer Gerhard Reisig remembered von Braun as a loyal and inclusive advocate for his staff, but he was prone to flights of fancy, requiring Dornberger to redirect his boundless energy: "Basically, they had the same ideas. But of course, von Braun was so interested in everything that he was a little bit in danger of going in a sideline, and Dornberger had to see that the whole thing came to bear, and so he had to call him back and say, 'Go on the main line and forget about your sidetrack, until after the war.' "[31] Eberhard Rees admired von Braun's "talent in dealing with people" and recalled the communal thrill of working in an exciting field like rocketry. "Peenemünde was for engineers a most interesting place."[32] Von Braun communi-

Peenemünde Army Research Institute, missile test site, spring 1941. *First row, from left to right*, Colonel Walter Dornberger (Dornberger was promoted to Major-General in June 1943); General Friedrich Olbricht (with Knight's Cross); Major Heinz Brandt; Wernher von Braun, age twenty-nine; and an unidentified man. General Olbricht played an important role in Operation Valkyrie, the failed July 1944 plot to assassinate Hitler and assume control of the regime. *Source:* Bundesarchiv, Bild 146-1978-Anh.024-03.

cated his enthusiasm and vision to the one who mattered most in the Third Reich—the Führer. Hitler was so impressed with von Braun that he insisted on personally signing the document promoting him to full professor.[33] For all of Dornberger's alliances, bargaining, and cunning, von Braun's slick presentations on the V-2's potential destructive capabilities were what won Hitler over and fast-tracked the project.[34]

Von Braun was a dreamer like Hitler, albeit not an ideologue, and he made others believe all things were possible. The V-2 was a weapon, of course, but the technology involved constituted a revolutionary breakthrough toward realizing von Braun's ultimate objective—space travel. He and Dornberger had to calibrate their sales pitch to a new audience at war's end. Detained in some comfort in the Bavarian Alps in June 1945, von Braun sat down to write a sweeping treatise on the brilliant future of rocketry thanks in no small part to himself: "We are convinced that a complete mastery of the art of rockets will change conditions in the world in much the same way as did the mastery of aeronautics and that this change will apply both to the civilian and the military aspects of their use. We know on the other hand from our past experience that a complete mastery of the art is only possible if large sums of money are expended on its development and that setbacks and sacrifices will occur, such as was the case in the development of aircraft."[35]

Von Braun seduced Dornberger, Hitler, American military officers, industrialists, presidents, and the American people with his dreams for nearly four decades. First, however, the proven team of Dornberger and von Braun had to reinvent itself and disassociate its work from the Third Reich's excesses. Writing in 1954, Dornberger lamented the fact that Hitler's "highly imaginative" mind "was unable to pass beyond the earth's atmosphere."[36] The man who kept von Braun grounded and focused on raining destruction on England was born again a dreamer. Those dreams came at a terrible human price, and the sacrifices to which von Braun alluded were measured in tens of thousands of slave laborers.

V-2 rockets at Peenemünde Army Research Institute, missile test site. V-2 on launch pad, V-2 on trailers, and V-2 launching. *Source:* Bundesarchiv, RH8II Bild-B0517-44.

Complicity in War Crimes

Nazi science enabled the worst atrocities committed by the Third Reich. Scientists and engineers, even those with few party affiliations or minimal ideological commitment, willingly lent their talents to the regime in exchange for professional advancement and the opportunity to continue their research with limitless funding or institutional support. Many in the scientific community supported the racial policies espoused by the regime, but even those uncomfortable with the extremist worldview preferred to serve the state than stand on the outside looking in. Consequently, scientists and engineers, specifically the V-2 specialists and others involved in aeronautics, participated in several areas of criminality. First, slave labor built the panoply of wonder weapons for the Third Reich. The SS reacted to and interacted with the experts when supplying the bodies needed. Second, several scientists acquired through Paperclip had some involvement in the notorious medical experiments conducted on concentration camp victims.

The issue of scientists' complicity in war crimes is at the core of the historical controversy swirling around Paperclip. Jean Michel, a forced laborer who toiled at Dora, forces us to confront the uncomfortable fact that "men who were intimately involved with the creation and operations of the camp are today respected, venerated and admired." Moreover, the "unspeakable sum of suffering, misery and death" not only helped manufacture missiles for Hitler's thousand-year Reich, "but . . . after the Russians and the Americans had shamelessly scooped up the scientists . . . made possible the conquest of space."[37] American military officers and employers were all too willing to excuse, minimize, and eventually fabricate the Paperclippers' backgrounds to expedite their travel and ensure long-term exploitation inside the United States. When the truth inevitably came to light, sometimes within months of a scientist's recruitment, the American public had already lost interest. After the government abandoned an earnest denazification policy in 1946, Paperclip accelerated, and a steady flow of recruits arrived at several installations in the United States. Only decades later was the legacy of Paperclip challenged by the significant paper trail incriminating some of Paperclip's most famous recruits.

On the night of August 17, 1943, 596 Royal Air Force bombers targeted the Peenemünde complex, killing hundreds of slave laborers and some scientists.[38] Operation Hydra also severely incapacitated the V-2 program's infrastructure and production schedule.[39] American bombers from the Eighth Air Force continued the onslaught during the day. The bombings forced the V-2 program to go underground, and with it, tens of thousands of slave laborers supplied by concentration camps and an endless supply of prisoners of war (POWs). After years of conspiring to burglarize secret weapons development, especially upon learning of the V-2's potential and Hitler's admiration for the project, Operation Hydra provided the SS with the pretext it needed to seize control of the missile program. SS involvement in the V-2 and related projects necessarily involved scientists and engineers in decisions and processes reliant on criminal acts, principally slave labor.

Shortly after the air raids, Hitler acquiesced and ordered SS chief Heinrich Himmler to use forced labor to safeguard the V-2 and increase production.[40] The SS expanded a facility in a mountain called Kohnstein, near Nordhausen, in the province of Thuringia. The SS and Albert Speer's Ministry of Armaments together created the Mittelwerk Company, comprising several SS officials to oversee construction and manage the project.[41] After first leveraging labor from neighboring Buchenwald to build the underground complex and then work on the assembly line, the SS established Mittelbau-Dora in summer 1944 as a separate camp serviced by a network of forty smaller camps.[42] The Dora camp predated the evacuation of Peenemünde, supplying twenty thousand prisoners for the production of Junkers aircraft and the *Vergeltungswaffe* 1 (retribution weapon), or V-1, "flying bomb" produced by the Luftwaffe.[43] The Mittelbau complex provided approximately sixty thousand laborers for the V-2 and V-1 from August 1943 to April 1945. In that time twenty thousand died from starvation, sickness, injuries, mistreatment, and executions.[44] Andrè Sellier, an inmate at Dora and a historian of the camp, described being "caught in a power struggle between the SS and the civil and military technocrats: in other words, between Arms Minister Albert Speer and his engineers, Wehrmacht military specialists, and the team of rocket specialists behind von Braun." The systematic use of concentration camp labor, to which none of the interested parties objected, aided the SS in its ambition to control the Reich's war industry.[45] According to recently discovered documents, the SS went so far as to test the V-2 on German villages in Pomerania and blame the destruction on Allied bombers.[46] Such callous depravity on the part of the SS only minimized the Paperclippers' "lesser crimes" of consent or indifference during the Paperclip deliberations.

If the SS became the alibi for a nation, exonerating millions of "ordinary Germans" from culpability for the crimes of the Holocaust, it certainly provided hundreds of Paperclippers a plausible explanation for their own actions working under an ever-expanding SS universe.[47] The US military was more than willing to absolve valuable Paperclip recruits, even if they held rank in the SS, a clear violation of the United States' own denazification policy. Of the many villains from which to choose in the Mittelwerk enterprise, SS Brigadier General Hans Kammler became the obvious scapegoat for building the tunnels with the same alacrity and zeal as he did the gas chambers at Auschwitz.[48] As head of Office C, the construction branch of the SS Economic and Administrative Office, Kammler oversaw mass production of the V-1 and V-2 for Himmler.[49] "Pay no attention to the human victims," he apparently told his staff. "[T]he work must proceed and be finished in the shortest possible time."[50] The SS assumed greater responsibility over Mittelwerk, and Kammler replaced Dornberger as the guiding force in planning and production.[51] Kammler aided the embattled Paperclippers' cause of absolution by supposedly committing suicide on May 9, 1945, thereby escaping prosecution and the opportunity to incriminate the civilian scientists and engineers he commanded.[52] Gerhard Reisig recalled Kammler's hostility toward Dornberger and especially von Braun, whom Kammler believed was a young, unreliable upstart "absolutely incapable of directing such a project."[53] Herbert Axster, a key figure for the rocket team during and after the war, described Kammler as "the worst man I'd ever met in my life."[54] Yet, not everyone discounted Kammler as a war criminal. Eberhard Rees recalled Kammler as "a man whom you could work quite well, and he was quite understanding." Rees, challenging the conventional wisdom concerning the SS general, claimed that although Kammler was enormously powerful and strict, "we had no trouble with Kammler whatsoever."[55] These contradictory opinions about a notorious war criminal reveal more about the Paperclippers' perceptions of their experience at Mittelwerk than about Kammler's legacy. Kammler delivered tens of thousands of laborers and did more to mass produce the V-2 than any previous official. Some Paperclippers failed to balance this accomplishment against the human cost. It simply did not concern them; at least that is how they explained the experience to Allied interrogators. Kammler's year at Mittelwerk coincided with the presence there of several important Paperclip recruits, Arthur Rudolph and Wernher von Braun most importantly.

More people died building the V-2 rocket than were killed by it as a weapon, a truth underscoring the brutality of the regime and the ineffectiveness of the V-2 campaign in the final months of the war.[56] Delving into the specific crimes associated with Mittelbau-Dora is important for establishing the complicity of key Paperclip scientists in war crimes. It also helps assess the reliability of the information US authorities used to evaluate Paperclip recruits for long-term employment and immigration to the United States. Furthermore, the contested legacy of Paperclip originates with questions of guilt and responsibility for the scientists and, for the US military, the question of "what did you know and when did you know it?" US efforts to expedite investigations and to minimize or fabricate intelligence information relating to Paperclippers' involvement in war crimes provoked bitter internecine disputes within the national security bureaucracy. As one arm of the government sought to investigate and prosecute individuals targeted as valuable assets by another, the military and some private industries worked to conceal the paper trail linking Paperclippers to war crimes. Ironically, military officers were the first to discover the crimes the military had worked diligently to conceal.

The US Twelfth Army liberated Mittelbau-Dora in April 1945 and quickly assessed the scope of the crimes committed from the few remaining survivors. Former New York governor Hugh Carey, a soldier and witness to the grisly discovery, remembered seeing "six thousand bodies stacked like cordwood, limbs separated, abandoned." The thousand survivors were clearly left to starve to death.[57] A report prepared by the Army Judge Advocate General detailed the horrific conditions for prisoners, including sleeping in the tunnels, contracting respiratory diseases, experiencing persistent abuse, and living in terror of random executions. The SS, the report cites, "shot laborers upon the slightest pretext. Six prisoners were shot one day because they had left their place of work to go to the water closet. Beatings were frequently administered upon the slightest provocation by SS guards and Kapos. One instance of a prisoner's failure to tip his hat to an SS man was sufficient justification for his being beaten." The report noted several instances of mass hangings of prisoners suspected of "sabotage" attended by the entire personnel of the camp—seventy-five prisoners on March 10, 1945; eighty in February 1945.[58] While workers passed along shoddy work and acted out in small ways, survivors recall how difficult it was to truly sabotage the process with so many nationalities involved.[59] Yves Béon, a French Resistance prisoner, witnessed "thousands of hangings" to the point that he "didn't pay attention."[60] Allied authorities interpreted the collective behavior of German civilian and SS staff and the relentless pace of production as "a common design" engendering mass starvation, beatings, torture, and executions. The "callous disregard" for living conditions and medical needs, prosecutors believed, "constituted criminal behavior." Of the nineteen defendants in the case involving Dora, most were SS personnel. The only civilian scientist named, general director

Prisoners' bodies lined the street of Nordhausen concentration camp in April 1945, when US forces liberated this and other Mittelbau-Dora camps, the source of slave labor for the V-1 and V-2 weapons. *Source:* United States Holocaust Memorial Museum, courtesy of Michael Mumma, 26811.

of Mittelwerk and Paperclip recruit Georg Rickhey, was acquitted.[61] Despite escaping immediate legal responsibility for the criminal enterprise of Mittelwerk, civilian scientists answered to charges in subsequent trials and investigations prompted by historians and government investigators seeking accountability.

Although the scientists, technicians, and engineers at Dora had less blood on their hands, Dora survivor Jean Michel deemed them "the aristocracy of evil." Michel did not believe personalities like von Braun and Dornberger "personally brutalized deportees" and acknowledges they "would have preferred to see their marvelous missiles made in more civilized factories and by a better treated work-force." Michel claims, however, that the "Peenemünde scientists' knew perfectly well what crimes were being perpetrated at Dora. Many fellow prisoners saw them in the tunnel and in the workshops."[62] Yves Béon depicts the scientists' presence at Dora in vivid detail: "Civil engineers,

Outside Boelke Kaserne, the central barracks of Nordhausen concentration camp, upon liberation in April 1945. Approximately 20,000 prisoners died from abuse, starvation, and horrible conditions inside the underground Mittelwerk factory. *Source:* United States Holocaust Memorial Museum, courtesy of Michael Mumma, 26816.

indifferent to the pitiful condition of the prisoners near them, are continually measuring the galleries according to the plans they carry. They move about, climbing the piles of rubble, going around machines and reels of cable, past turning concrete mixers, but never looking at the tattered men around them, nor even hearing the shouts, the vicious clubbings, or screams of pain. Quietly, they indicate location and points desired for machines, for junctions, for joints and fixing points for the electric and pneumatic air ducts."[63] Despite comprehensive evidence and testimony emanating from survivors, prosecutors, and historical scholarship, the horrors of Mittelbau-Dora were eclipsed by Paperclip's increasing value to the national security state and the near invisibility of the camp after falling under Soviet control and eventually that of East Germany.

Several scientists who escaped scrutiny concerning their actions in Mittelwerk, in some cases for decades, answered for them later. Peter Wegener helped develop the hypersonic wind tunnel at Peenemünde and later worked at the Naval Ordinance Laboratory and Yale University. In his 1996 memoir Wegener recalled visiting Mittelwerk during the war and feeling "the only decent thing to do would be to rip off my uniform, put on a striped suit, and join the prisoners that all of us had put into such an inhuman situation."[64] While it is commendable to acknowledge this reality decades later, especially for someone tangentially involved in the criminal decisions relating to labor, Wegener is the exception. Arthur Rudolph and Wernher von Braun were invaluable Paperclip acquisitions. Consequently, their relationship to the Nazi regime's war crimes prompted great interest by advocates and detractors alike.

Arthur Rudolph

The case of Arthur Rudolph is central to the questions surrounding the legacy of Paperclip. Rudolph was a talented engineer and administrator throughout his forty-year career, beginning in Peenemünde and the tunnels of Mittelwerk, followed by his work at US Army facilities in Texas and Alabama, and concluding with a storied career at NASA as project director for the Saturn V rocket and a contributor to the Apollo space program. Rudolph's record of success, along with the successes of dozens of other key recruits, seemed to vindicate Paperclip. His career in the Third Reich was deemed irrelevant and, for the most part, unexamined by army investigators. As the factory director responsible for establishing the production line for the V-2 at Mittelwerk, Rudolph worked closely with the SS to acquire slave labor and meet the fantastically high production goals.[65] Rudolph agreed decades later that the SS "was a rival," but "I don't say that everything that the SS did was bad." Rudolph admired their research on jet engines and appreciated SS protection of valuable equipment at Mittelwerk.[66] Rudolph exploited the SS "rent-a-slave service," although it was not officially under his purview.[67] With so much attention focused on Hans Kammler and other senior SS officials, Rudolph's name was conveniently buried deeper in the archives and, like most Paperclip recruits, Rudolph benefited from an expedited security investigation. It was not until Eli Rosenbaum of the OSI read accounts of Mittelbau-Dora that Rudolph's case was reexamined, and with it the questionable practices of responsible agencies involved in Paperclip. The investigation resulted in Rudolph renouncing his citizenship and returning to West Germany in 1984.

First, it is important to establish Rudolph's function in the murderous production center at Nordhausen. Rudolph depicted himself in interrogations as a hapless engineer reacting to SS demands and pressure, but he instigated the process to utilize slave labor at both Peenemünde and Mittelwerk after touring an SS factory at the Sachsenhausen concentration camp. Civilian experts like Rudolph determined labor needs, and the SS provided an unlimited supply. As early as February 1943, before Peenemünde began production of the V-2, Rudolph planned to rely on "Russians" acquired from nearby POW camps for labor. Dornberger agreed, dismissing the inmates as "murderers, thieves, criminals."[68] Rudolph obviously studied the problem and assessed the risk, noting the strengths of the SS policy: "[T]he mixing together of nationalities has the advantage of limiting the formation of secret resistance groups."[69] Andrè Sellier rightly describes Rudolph's revelations concerning the value of slave labor as "the beginning of a process that would lead to the deaths of thousands of prisoners in Dora." The documents accurately portray Rudolph as "concerned exclusively with solving his labor problems, both qualitatively and quantitatively, in whatever way necessary."[70] In his June 1947 interrogation by the US Army, Rudolph deftly shifted responsibility to the SS, claiming his own living conditions and food rations were no different than the prisoners working in the tunnels, which, he maintained, he was not allowed to visit.[71] Rudolph recounted the hanging of twelve prisoners from a crane in fall 1944 and again claimed no responsibility, but eyewitness reports and further investigations, including his subsequent interviews with OSI investigators, contradict this initial claim.[72] In a 1989 interview, the exiled Rudolph continued to deny involvement with forced labor, claiming never to have seen "any people in striped suits in Peenemünde." If there were laborers working on construction, he states, they were "volunteers." Rudolph was steadfast in stating concentration camp inmates likely worked construction, but "certainly not in my plants."[73] The document trail is extensive thanks to historians like Michael Neufeld, who unearthed much of the evidence, including a June 2, 1943, memo in which Rudolph "presented a formal request for fourteen hundred concentration camp prisoners, subdivided by skill."[74] The Rudolph exposure fell short of full accountability, but it forced the nation to confront deeper questions concerning Paperclip and other postwar intelligence operations involving more sinister figures than Rudolph.

Wernher von Braun

Although Wernher von Braun rose to the rank of *Sturmbahnführer* (major) in the SS, by the war's end, his relationship to the organization was mostly utilitarian. Von Braun joined the organization at the behest of Dornberger, who recognized the growing power of the SS and their interest in rocketry. Von Braun, however, joined in fall 1933 ostensibly to avail himself of the SS horse riding school.[75] The duo tolerated the SS and obviously benefited from slave labor, but their relationship with the SS turned for the worse after the failed plot to kill Hitler in July 1944.[76] Von Braun equated SS help to "an excessively generous dose of 'liquid manure' that would kill the 'little flower' of the rocket program."[77] He recounted his strained relationship with the SS in a June 1947 affidavit explaining his membership in the National Socialist German Workers' Party (NSDAP) and SS. Heinrich Himmler had summoned von Braun to his headquarters in East Prussia in summer 1944 to begin luring the V-2 into the SS universe. "He [Himmler] said he realized the Army is 'full of red tape' and that he could imagine what a 'poor inventor like me' had to suffer from that kind of difficulties," von Braun wrote. Himmler promised von Braun an "open door to the Führer" and the ability to solve all the V-2's production problems. Von Braun and Dornberger held Himmler at arm's length until the Reichsführer changed tactics and arrested von Braun only eight weeks later for suspected "anti-Nazi utterances."[78] Among the charges, and they ring true, von Braun obsessed more about space travel than military necessity when it came to the V-2. Only after Albert Speer and Dornberger intervened did the SS release von Braun three weeks after his arrest. The arrest helped von Braun's postwar career, allowing him to play the reluctant civilian scientist dedicated to the future, not a willing accomplice. Von Braun may not have been a devoted party follower, but he bears some responsibility for the casualties wrought by the V-2, both innocent British civilians and the thousands of slave laborers who perished in the tunnels of Nordhausen.

Von Braun biographer Michael Neufeld is the authority on determining von Braun's actions and awareness regarding slave labor. Unlike Rudolph, von Braun was not transferred to Mittelwerk after the raids on Peenemünde. "The organizational separation of development and production," Neufeld wrote, "meant that von Braun bore no official responsibility for decisions to use forced and slave labor, and if he had any qualms, which is doubtful, he scarcely could have objected anyway. But he was well-informed about these decisions."[79] Still, von Braun exploited the pool of labor allocated to realize his dream project. In August 1944, von Braun visited the barracks at Dora and spoke with some prisoners with the requisite education and experience to assist with V-2 production. Charles Sadron, a French scientist working in Mittelwerk as a forced laborer, recalls meeting von Braun and his offer to work for him directly. Sadron refused but notes it was a polite exchange. Von Braun took it upon himself to place technically skilled prisoners in various positions.[80] Von Braun knew the commandants of Buchenwald and Dora personally and wrote an August 1944 letter acknowledging his use of slave labor: "I went . . . to Buchenwald . . . to seek out more qualified detainees. I have arranged their transfer to the Mittelwerk with Standartenführer Pister." Neufeld considers references in documents like this "the most damaging pieces of evidence yet" and concludes such a transfer of labor "in theory put him in violation of the Nuremberg standard applied to Albert Speer."[81]

Von Braun could not avoid his proximity to obvious war crimes indefinitely. In 1969, while America celebrated the heroic *Apollo 11* moon landing, von Braun reluctantly provided a witness statement, on behalf of an East German prosecution of three SS officers accused of war crimes at Mittelbau-Dora, in which von Braun admitted witnessing the underground sleeping arrangements. In a 1976 interview, shortly before his death, von Braun supplied more details: "The working conditions were absolutely horrible. I saw the Mittelwerk several times, once while these prisoners were blasting new tunnels in there and it was a pretty hellish environment."[82] Surely, von Braun knew as much, if not more, than his secretary Hannelore Bannasch, who regularly witnessed the conditions in the tunnels. Bannasch provided a sanitized and revisionist account of her experiences to the rocket team's sympathetic biographers and von Braun's colleagues and historians, Frederick Ordway and Mitchell Sharpe: "At first the laborers slept in the tunnels—Germans and foreigners alike. Because of the dampness, many died of pneumonia. Actually, as time went on we got to work quite well with the foreigners—it was a veritable melting pot. But they often fought among themselves. Remember, many had become prisoners for criminal and homosexual reasons as well as for their political and religious beliefs. We needed the laborers, so we tried not to mistreat them. It [Mittelwerk] was a top secret operation, so once you were in you stayed."[83] The notion that the laborers toiling underground were treated as well as Germans or, failing that, deserved their prisoner status is ludicrous and offensive, but the ambiguity surrounding von Braun's personal responsibility helps preserve his heroic stature among many Americans.

Medical Experiments and Testing

US employment of German scientists implicated in unethical and criminal experiments on concentration camp inmates provoked understandable outrage in government circles and among the broader public. German advances in aeronautical medicine as well as chemical and biological warfare involved cruel experiments on concentration camp prisoners. This book will not replicate the work of investigative journalists who exposed these events over years, but it is helpful to understand what information was available to military authorities before and during the recruitment process. Samuel Goudsmit, the director of the Alsos mission intended to assess Germany's nuclear program, recommended circulating documents incriminating German scientists to their Allied counterparts: "The ultimate purpose of this procedure is to acquaint American specialists with the activities of their German colleagues. This is expected to have a profound influence upon their attitude towards enemy scientists." Rather than revulsion, Tom Bower notes, such findings only piqued interest.[84] Allied investigators discovered evidence of war crimes during their work securing research and development centers across the collapsing Reich. In a conference between US, French, and British FIAT personnel, the team reported that "evidence showed that some German doctors and scientific research workers had engaged in unethical experiments on living human beings," specifically citing freezing experiments and "war gases" on concentration camp victims. The team decided gathering evidence fell under the FIAT mission and arranged to transfer files to criminal investigators.[85] IG Farben board member and chemist Fritz Ter Meer spent seven years in prison for his role in constructing the Monowitz concentration camp near Auschwitz and testing chemicals on some twenty-five thousand prisoners. FIAT interrogators asked Ter Meer if the experiments were justifiable: "He argued that no harm had been done to these KZ [concentration camp] inmates as they would have been killed anyway and were thus offered a chance of survival, the tests had humanitarian aims because they were to help others in the future."[86] Such specious reasoning resonated with American authorities as well, especially regarding aeronautical medicine.

Otto Ambros, another IG Farben official, experimented with nerve agents on the same population of inmates as Ter Meer. Ambros's pioneering work was of great interest to American interrogators, but his flagrant criminal behavior disqualified him from exploitation until his early release from prison in 1952.[87] FIAT investigators noted his qualifications and importance, citing positively that Ambros had been rewarded by Hitler personally and named by Albert Speer as "the key man in German chemical warfare production." Ambros was the "guiding power" behind tabun and sarin production, but he used POWs to produce the gases and "concentration camp prisoners for human experiments."[88] Amazingly, French authorities allowed Ambros to continue work at an IG Farben plant in Ludwigshafen knowing full well that American authorities sought his arrest and interrogation at the detention facility known as "Dustbin." Ralph Osborne, the commanding officer of FIAT, was incredulous: "This man is thought to be far too dangerous and undesirable to be left at liberty, let alone employed by the allied authorities."[89] While never contracted as a Paperclipper, Ambros enjoyed a lucrative career with American and German industry after his release.

Aeronautical medicine was one field of German science both highly desired by the US military and rife with charges of criminality. Hubertus Strughold was director of the Aviation Medical Research Institute for the Luftwaffe and played a crucial role in establishing a similar institute at Randolph Air Force Base in San Antonio, Texas. Immediately after the war, Strughold arranged for more than one hundred scientists to write summaries of their research and present the material to the Allies, although many were arrested soon after. Thirty-four of the scientists affiliated with Strughold accepted Paperclip contracts.[90] If von Braun helped humanity reach the stars, Strughold, the "father of space medicine," helped humans survive the journey. Unfortunately, Strughold's pioneering research is tainted by involvement in the horrific "cold experiments" conducted at Dachau and recent revelations that he experimented on epileptic children at a Berlin psychiatric hospital.[91] Strughold's case resembles Arthur Rudolph's and sparked similar debates on the legacy of Paperclip and medical ethics more generally. Several Paperclippers discussed in the book have similar charges in their background, including Strughold associate Konrad Schaeffer; Ernst Eckert, an aeronautical engineer also implicated in cold-resistance tests; and Walter Schreiber, an epidemiologist and biological warfare expert connected to experiments on concentration camp inmates. Unlike Strughold and Rudolph, Schreiber was forced to leave the United States soon after accepting a Paperclip contract in 1951. These cases highlight the imprecise and contested process of targeting, evaluating, and employing Paperclip scientists. The US armed services and intelligence services, not unlike the Third Reich, comprised competing fiefdoms battling over limited resources and pursuing contradictory agendas under the nebulous umbrella of "national security."

Vetting "Unrepentant Nazis"

On June 8, 1945, Walter Jessel was assigned to a small Counter-Intelligence Corps (CIC) detachment to "screen" members of Wernher von Braun's rocket team, detained outside Garmisch-Partenkirchen in upper Bavaria. Jessel had explicit orders from Supreme Headquarters Allied Expeditionary Force (SHAEF) to sort out, in his words, "Nazi hangers-on and enforcers from technical staff in order to bring the latter to the US."[92] Jessel and his colleagues faced a difficult task distinguishing between the invaluable brains behind Hitler's wonder weapons and those who were either expendable or so tainted by the regime's criminality as to preclude contractual employment of any kind. As forthright as Jessel's screening report reads, his diary entries from that week in June are even more frank: "The team consists of rocket enthusiasts, engineering college graduates, professors, all unrepentant Nazis aware of their bargaining power with the Americans." Jessel notes how Wehrmacht personnel attached to the team understood "that their chances of going to the US are smaller than those of technicians. To improve these chances, they sing."[93]

Jessel reserved his harshest assessment for General Walter Dornberger, whom he described as incredibly arrogant and condescending, lecturing Jessel about how war fuels civilization and rocketry is the next great leap forward

Second Lieutenant Walter Jessel, June 1945. Jessel, a German Jewish émigré, was among the first to interrogate members of the rocket team. *Source:* Courtesy of Alfred Jessel.

on par with the invention of the wheel. "His group's ambition was to develop a weapon with which to dominate the world," Jessel writes. "They are mercenaries who want to sell their weapon. Their country is defeated, hence their only chance is to go on doing the same business for someone else." Dornberger courted the British as well, although they were more interested in imprisoning him for building the V-2 than rewarding him with freedom and a lucrative contract.[94] Dornberger told British interrogators the V-2 could be a weapon, transport mail and passengers, or carry "atom bombs upteen thousands of miles." "I do not know which direction development will proceed," stated Dornberger, "Upon that depends the selection of the people to be employed and the type of installation required."[95] Von Braun never shied away from military uses for his technology, but Dornberger recognized better than Jessel how prominently the specter of the Soviet Union loomed over US military thinking. Jessel was not unimpressed with Dornberger: "[W]hen you get off his obnoxious philosophy, [*sic*] is better at explaining engineering developments to a layman than anyone I ever saw. He'll charm his new US bosses."[96] Jessel's skepticism was communicated up the chain of command, but it failed to sway decision makers anxious to retain the rocket team's services. However, Jessel managed to write a revealing profile of prominent Paperclippers in just a week's time.

Jessel divided the team into three groups: the "early developers of rocket ideas," the technicians "from engineering schools and industry" and army personnel who were transferred to Peenemünde in 1941. "The second group is the largest, and ideologically the least sympathetic," Jessel suggests, noting most were party members and attended "Nazified" technical schools which churned out inferior engineers dependent on the regime for employment. Jessel acknowledges that "the number of Nazi fanatics is not very large" among the detainees, especially since they lost the war and were therefore denied credit for developing revolutionary technology. Nor is there any sense of sharing "Germany's guilt and responsibility." Jessel was most troubled by the team's mercenary mentality and disingenuous attempts to stoke fears of the Soviet Union: "Almost to a man these people are convinced that war between the US and Russia is around the corner. They shake their heads in amazement and some contempt at our political ignorance and are impatient at our slowness in recognizing the true savior of Western Civilization from Asia's hordes. Which does not prevent them from playing with the idea of selling out to Asia's hordes if such recognition is not soon extended." The Soviet peril secured the scientists a meaningful future and cleansed a murky past.[97] V-2 scientist Herbert Wagner told his interrogators, "We had realized, long before anyone else, what a menace the Soviet Union was to Western civilization and culture. And that is why we helped Hitler."[98] The CIC interrogators believed the team depended on Dornberger and von Braun for their livelihood and noted how individual Germans expressed little confidence in "their own ability, technical or personal, to make their own way in Germany or elsewhere."[99] This fact, Jessel notes, provided von Braun and Dornberger excessive sway over recruitment. The chief scientific adviser to FIAT, Henry Robertson, resented someone who "is not a first rate technician" like Dornberger. "I am convinced that Dornberger is a most dangerous man," Robertson opined, "and that he should in any case be shorn of all influence over and even prevented to have contact with his former Peenemünde subordinates."[100] Another official quipped it might be better to "trade him [Dornberger] to the Russians for a dish of caviar."[101] Not for the first time, scientific advisers were overruled by officers bent on achieving the next breakthrough.

The rocket team was interrogated in preparation for Operation Overcast, which evolved into the more robust Paperclip program months later. The CIC team cultivated several informants hostile to the control exercised by a clique comprised of Dornberger, von Braun, Herbert Axster, and a few others from Peenemünde. According to the informants, the clique issued directives to fellow detainees, ordering them to "[d]ivulge no more than required to arouse technical and scientific curiosity in US interrogators" and to "[a]void going into technical detail and giving any trade secrets which might make the US project unnecessary." When pressed, the informants reported, they were told to blame "[t]he absence of documentary material without which details could not be worked out or recalled." In some cases, members of the team buried documents for this purpose. Another tactic involved referring interrogators to other scientists "in order to enlarge the number of personnel 'indispensable' to the US project." The clique urged detainees "take notes immediately after interrogations, so that consistency in future interrogations might be assured." If someone balked or contradicted the directives, the detainee could expect disciplinary measures ranging from reduced rations to disqualification as a potential Overcast acquisition. The informants warned that the clique succeeded in convincing the US Army that Overcast recruits should be knowledgeable in a specialized field, subordinate to the clique's leadership, and exhibit the proper esprit de corps.[102] The informants related the clique's opinion that a successful V-2 test launch and judicious information sharing with the Americans "served merely as an appetizer." Von Braun, especially, continued to exert strict control over his team throughout his career in the US Army and NASA, usually without interference from his employers.

The CIC's informants predicted the rocket team's intentions over the long term, specifically linking past behavior to future performance. "[T]he rocket developers became outstanding promoters," one report read, "They are none too happy about US Army supervision of the project. They hope that during the year they will have opportunities to contact US industrialists who would then have to be sold on taking over the sponsorship." Most informants agreed, however, that von Braun would get results "regardless of his employer" provided he managed the team the way he wanted.[103] Whether team members resented US military oversight or not is questionable, but they certainly sold interrogators on their grandiose visions. Intelligence agent Charles Stewart concisely summarized the clique's reasoning during the Third Reich's final days: "They had selected the Americans, as they were favorably disposed to this country generally and also because this country was the most able to provide the resources for interplanetary travel."[104] Interrogators gained insight into the internal dynamics of the rocket team and disclosed the cynical manner in which its members conducted business with their captors, but much of this information was ignored.

The informants' revelations combined with Jessel's evaluation of the rocket team's dishonesty and Machiavellian worldview raised obvious questions about security, in addition to the quality of information peddled by the rocket team. "Considering the security question only," Jessel writes, "the very existence of the group[,] be it in Germany or the US, constitutes a danger. It leaves Dornberger and von Braun in a position to bargain, play the US against her allies, control research results, the output of individuals, and their political attitudes." Jessel warned the team could "exercise pressure and attempt blackmail" if their research progressed. "Their position toward US authorities might somewhat resemble their relations with the SS after July 20, 1944. The SS appears to have been rather unsuccessful." In other words, the rocket team served itself better than any patron. Jessel concludes with words of advice, which went unheeded: "Individuals of this group, if separated from it and exposed to American methods and ways of life, may under close control develop more human traits. . . . Individual screening can only suggest this possibility. Security clearance of the group as such is an obvious absurdity."[105] Jessel's report is a valuable historical document, but it raised few alarms inside the US military. The prize was too rich, and despite their obvious manipulation, the rocket team's dire predictions about the Soviet Union's advances and nefarious intentions mirrored the military's own thinking.

As Allied intelligence teams scoured German territory in the final months of the war in a frantic search for advanced research equipment and documents, the name Werner Osenberg became synonymous with a master list of the best and brightest minds in the German military-industrial-university complex. Osenberg, a trained mechanical engineer and senior SS Gestapo officer, was tapped in June 1943 to supervise the Planning Office of the Reich Research Council. During his assignment fulfilling Hitler's June 1942 decree requiring "leading men of science" to work on military projects, Osenberg compiled a list of approximately fifteen thousand names.[106] In March 1945, a Polish lab technician discovered portions of the list in an unflushed toilet, from which it promptly found its way to the Combined Intelligence Objectives Subcommittee (CIOS) organized by British and American (Anglo-American for short) intelligence services. Osenberg voluntarily surrendered the complete list along with summaries of the scientists' expertise and a trove of documents, such as research contracts and the status of "special weapons" like the V-2.[107] US Army major Robert Staver relied on the Osenberg file to create the military's black list of targets used to direct Project Paperclip recruiting efforts for the next decade.[108] The Osenberg list led Allied forces to von Braun's rocket team, aeronautical medicine specialists, and various experts in jet engines, chemical and biological warfare, nuclear physics, and synthetic fuels, just to name a few.

Statistics concerning Paperclip scientists acquired between 1946 and 1952, particularly relating to age and career placement inside the United States, suggest recruiters valued youth and the military applicability of their expertise. In 1946 only 129 specialists worked in the United States, but that number rose to between 492 and 555 during the years 1948 to 1952.[109] Of the approximately 525 Paperclippers residing in the United States during this period, forty percent worked for the air force, thirty-four percent for the army, sixteen percent for the navy, and ten percent for the Department of Commerce.[110] Burghard Ciesla cataloged the birthdates of hundreds of Paperclippers and determined that the vast majority were under forty years old in 1945. Two hundred scientists, the largest grouping, were born between 1900 and 1909, followed by 159 between 1910 and 1919. Sixty-two specialists were born before 1900.[111] William Bainbridge conducted a more specialized survey of the rocket team, acquiring the birth dates for 26 of 118 who arrived in the United States in 1945 and 1946.

The mean year is 1912, Wernher von Braun's birth year. Most of the rocket team belonged to the same generation. "In 1942, when the first V-2 was successfully launched," Bainbridge determined, "they averaged just thirty years old."[112] Von Braun recruited most of the team and tended to seek out young and ambitious engineers like himself.

In April 1948, the JIOA collated data on hundreds of Paperclip recruits and determined "that the majority of German scientists were members of either the Nazi party or one or more of its affiliates. These investigations disclose further that with very few exceptions, such membership was due to exigencies which influenced the lives of every citizen of Germany at that time."[113] Singularly interested in facilitating immigration for the largest possible number of targeted scientists and technicians to the United States, the JIOA consistently diffused this assessment throughout the bureaucracy. Discerning the truth behind party membership is difficult because many in the "generation of the unbound," the von Braun cohort especially, relied on government patronage for advancement. Furthermore, universities and professional associations all but required party membership to access scholarships, grants, and eventually employment in the postcoordination Third Reich. Wind tunnel expert Peter Wegener recalls his security interview with the Americans and assumed most of his fellow Paperclip recruits "had been in the party and related organizations. Where would the line be drawn?"[114] Despite a united effort on the part of the JIOA and the scientists they coveted to minimize the record, security investigations conducted by military and civilian agencies provide considerable details about Paperclippers' backgrounds.

The US Army, the primary investigative agency for Paperclip in its initial phase, defined an "ardent Nazi" as someone who joined the party before Hitler came to power in January 1933, occupied a leadership position in the party or affiliate, was convicted in a denazification court, or was accused or convicted of a war crime.[115] Investigators worked within the confines of the denazification program announced by the European Command (EUCOM) in July 1945 when evaluating Paperclip recruits. Ardent Nazis were problematic but not always excluded from consideration, especially if the recruit had expertise in desirable fields. Respondents to the infamous questionnaire (*Meldebogen*) detailing party activities fell into one of five categories: Major offenders (Class I); offenders (Class II); lesser offenders (Class III); followers or nominal Nazis (Class IV); and persons exonerated (Class V).[116] Military and civilian authorities generally adhered to prohibiting Class I and Class II offenders from Project Paperclip. In May 1948, investigators assigned to EUCOM outlined their criteria for rendering a positive or negative security evaluation:

> Each case must be judged as an individual one and a final evaluation reached only after apparent motives and the degree of political fanaticism have been determined. Many scientists were honorary SS members and, likewise, many received party awards for the contribution they made to the German war effort through their scientific achievements. Some were members of the SA [Sturmabteilung] while in college, a requisite for entrance into the college. There were many other scientists who joined the NSDAP for reasons of opportunity within their scientific field or who did so in order to escape Gestapo and party harassing.[117]

Normally SS membership or unassailable proof of ideological sympathy with the Nazi regime disqualified a candidate from any substantive employment inside Germany, let alone a generous contract in the United States. However, the JIOA not only ignored negative security reports, but often altered the offending documents to deceive other offices in the government.

The political climate at Peenemünde during the war is difficult to assess given its physical isolation and mostly youthful staff. Michael Petersen notes that party membership among the rocket team was "higher than average in Nazi Germany," as evidenced by a sample of eighty-four personnel files, but this fact alone does not necessarily "show a general affinity for National Socialism" among the young scientists given the level of party involvement in institutions of higher education.[118] Eberhard Rees, an engineer at Peenemünde, diminished the role of the party at the complex, perhaps out of self-interest and the desire to preserve a positive legacy, but his recollections correspond with other evidence. Rees remembered an engineer charged with running party meetings: "I knew him, real well because he worked with me and once in a while[.] [H]e said, 'Well, I have to have one of those meetings again,' but he was never a great Nazi. Peenemünde was never a big Nazi place."[119] This is a matter of perspective, but it is true that ideologues and war criminals coexisted alongside idealistic scientists, including some fiercely anti-Nazi individuals, at Peenemünde and at other scientific enterprises across the Reich. These divisions manifested themselves once Paperclippers arrived in the United States as well.

Every category of offender outlined in the denazification laws participated in the Paperclip program in one form or another. My own survey of approximately

130 personnel files belonging to Paperclip scientists assigned to different projects suggests diversity of background and ideological commitment as well as a pattern of methodical deception and synchronization of responses, confirming Walter Jessel's initial observations in June 1945. In most cases the JIOA and FBI investigated the same pool of applicants, producing independent assessments. Both are informative, but the FBI tended to be more thorough and inquisitive. For example, the bureau regularly informed the JIOA that it considered "any persons having connections with the Nazi party as a definite security threat to the internal security of the US."[120] This policy obviously changed over time. Detailing the backgrounds of every scientist here is prohibitive given space limitations, and this is not the sole instance in which I discuss scientists' backgrounds, but the following case studies highlight the politics of some prominent Paperclippers and the challenges facing agencies and historians charged with discerning the truth.

Magnus von Braun

On May 3, 1945, Wernher von Braun dispatched his younger brother, Magnus, on a dangerous mission to contact American forces and arrange for the surrender of the assembled rocket team. A trained chemical engineer, Magnus worked as his brother's personal assistant and constructed gyroscopes for the V-2 at Mittelwerk. The two traveled together to the United States, where Magnus, understandably, lived in the shadow of his lionized brother. Magnus also left a significantly more negative impression than his brother did in terms of ideological fervor. CIC agents assigned to Fort Bliss, Texas, where the rocket team would first conduct tests in the United States, described Magnus as a "dangerous German Nazi" who is "a worse threat to security than a half a dozen discredited SS Generals."[121] Michael Neufeld speculates that the younger Magnus received a "heavier dose of National Socialist indoctrination" as a Hitler Youth leader and active member of the National Socialist German Students League.[122] Magnus von Braun's JIOA dossier tells a different story, concluding he was "not a war criminal, not an ardent Nazi and not likely a security threat."[123] Magnus considered his memberships "obligatory" and noted they were "null and void when I was drafted to the Luftwaffe in 1940."[124] Major James P. Hamill, the principal reference listed in Paperclip scientists' JIOA dossiers, predictably had nothing but positive things to say about his subjects. Hamill acknowledged Magnus was "[h]andicapped by the reputation of his older brother, nevertheless a capable engineer in his own right. Conscientious and quiet."[125] Magnus von Braun was not so diligent, however, when he stole a platinum bar from the base and tried to sell it to an El Paso jeweler for one hundred dollars. The army requested that no action be taken "for security reasons and possible adverse publicity which might affect the long range objectives of the project on which the group of Germans was employed."[126] According to Hamill, "when the matter was brought to the attention of Wernher von Braun . . . he had administered a severe physical beating to his brother."[127]

Wernher von Braun

Despite Wernher von Braun's importance to the Paperclip program, there were reasons to distrust the celebrated creator of the V-2 as Walter Jessel had indicated. Members of the rocket team supposedly buried thousands of documents during their flight to Bavaria, but not all were recovered.[128] The CIC determined that von Braun and Herbert Axster were "chiefly responsible for concealing from the US authorities the existence of these caches and possibly removing a good part of the documents which have not yet been located." The report filed on September 27, 1947, accuses the two of "camouflage talk" and "decoy location sketches." "Both von Braun and Axster must stand convicted of lack of honesty in their dealings with their US employers as late as July 1947," the report reads. The CIC accused von Braun of dishonesty and continued subterfuge, evidenced by his coaching other scientists to lie. "This may explain the uniformity of their answers," the CIC concludes.[129] The CIC report reflects frustration at wasted resources and repeated failures thanks to false leads. It is surprising to read such negative reporting about von Braun two years into this Paperclip tenure. The CIC report further indicates that not all responsible authorities in the military were enamored with the rocket team.

Von Braun's party membership raises a similar issue related to honesty. Von Braun applied to join the NSDAP in 1937 and was given a party number the same year, but he told US authorities that he "was officially demanded to join the National Socialist party" in 1939. He claimed that "my refusal to join the party would have meant that I would have to abandon the work of my life," and that his membership "did not involve any political activity."[130] Michael Neufeld believes von Braun either lied about the date or was subconsciously attempting to cleanse his record. Von Braun was undoubtedly "mildly enthusiastic" about the Hitler dictatorship for the potential to pursue rocketry, but he apparently expressed no strong political ideas or antisemitic beliefs. He was promoted consistently through the SS, however, and never ran afoul of the organization until Himmler's August 1944 ploy to incorporate the V-2 program into the SS empire.[131] Regardless, von Braun's past scarcely mattered to his US employers. In February 1947, months before the scathing CIC report, the US Army Ordnance Department praised von Braun's cooperation, attitude, and most importantly, his results: "He would be most difficult to replace if not impossible." Von Braun's stated reasons for immigrating to the united States are virtually identical to dozens of other responses from members of the rocket team, presumably because they coordinated responses: "I want to devote my future work to the progress and strength of the Western Civilisation and I consider the United States of America their bulwark."[132] Von Braun's "technocratic amorality," as Neufeld terms it, served him well on both sides of the Atlantic.[133]

Arthur Rudolph

As with Magnus von Braun, the military's initial assessment of Arthur Rudolph in 1945 stands in stark contrast to subsequent evaluations. The first report read: "100% NAZI, dangerous type, security threat . . . !! Suggest internment."[134] Aside from Rudolph's intimate involvement in slave labor and executions, his history with the NSDAP alone raised troublesome questions. First, Rudolph had joined the party in 1931, indicating he was less an opportunist than a true believer. He explained his membership in terms most Germans attempting to survive the postwar era could understand: "After 1930 the economical situation in Germany became so serious that it appeared to me to be headed for catastrophe. The great amount of unemployment caused expansion of the national socialistic and communistic parties. Frightened that the latter one would become the government, I joined the NSDAP (a legally registered society) to help, I believed, in the preservation of the western culture."[135] Rudolph recalled in a 1989 interview, "When people said 'Heil Moscow'—then I preferred the Heil Hitler."[136] Rudolph consistently expressed his disappointment in Hitler for failing to deliver on his promises and claimed to reject the Nazi racial worldview. Like most Paperclip prospects, Rudolph apologized for his tainted past and pledged loyalty to the United States in early interrogations. The JIOA, in what would become a pattern in handling Paperclip paperwork, altered a word or two in the original Office of Military Government, United States (OMGUS), security evaluation. In March 4, 1947, OMGUS declared that Rudolph was "not a war criminal and was an ardent Nazi." The September 27, 1948, version of the report stated that Rudolph was "not a war criminal, was *not* an ardent Nazi, and in the opinion of the military governor, OMGUS, is not likely to become a security threat to the US."[137]

Later CIC reports, including interviews with coworkers, continued to generate evidence of Rudolph's ideological toxicity. In April 1953, Bruno Helm, a coworker of Rudolph's at Redstone Arsenal, characterized him as "a loyal member of the NSDAP," who earned "the reputation of being a person who in his enthusiasm for the Nazi regime could be dangerous to a fellow employee who did not guard his language." The agent conducting the interview believed Helm was motivated by jealousy, but there was more than enough evidence to justify Rudolph's involuntary return to Germany decades later.[138] Rudolph and Georg Rickhey, the general director of the Mittelwerk Company, escaped responsibility for war crimes by fulfilling roles as "technical personnel."[139] Rudolph enjoyed a forty-year career in the United States, however; Rickhey, just one.

Georg Rickhey

Georg Rickhey was extended a Paperclip contract in May 1946 and worked alongside dozens of other Germans at Wright Field, Ohio. Rickhey returned to Germany to stand trial a year later and was acquitted of charges, but the indictment alone was enough to disqualify him from further employment with the US military. Rickhey's expertise in underground facilities, which included constructing the V-2 factories at Nordhausen and Hitler's Berlin bunker, made him an obvious target for Paperclip recruiters. The army offered Rickhey an additional five-year contract in April 1947, just one month before the war crimes branch of the OMGUS Civil Affairs Division named him "a principal perpetrator in the Nordhausen Concentration camp case."[140] Rickhey attempted to reclaim his position after the acquittal, claiming the "the accusations were only based upon untrue evidences given by communists and the SS-Gestapo agents." Rickhey also blamed the Russians directly for undermining his contributions to American security.[141] The JIOA's embarrassment over the Rickhey revelation, however, precluded his reinstatement and figured into the internal struggle between the agency and the State Department, which had lacked confidence in the military's screening process from the outset. Furthermore, a colleague of Rickhey's at Wright Field and at Mittelwerk denounced Rickhey for their Nazi pasts. Fellow Paperclipper Hermann Nehlsen reportedly claimed Rickhey was "a strong Nazi party member, where in 1944 twelve foreign workers were simultaneously hanged by being strung up on a cross and raised by a crane in the presence of the group of workers. One of the group who acted as observer asserted that Dr. Rickhey was the chief investigator for the execution."[142] The claims mirrored the war crimes charges Rickhey faced, but his Paperclip career was over regardless of the outcome.

Albert Patin

Hermann Nehlsen, whose own dossier identified him as a "consistent opponent of National Socialism," also singled out Albert Patin, an industrialist and expert in aeronautical patents, for being "an old member of the NSDAP, a SA Standartenfuehrer . . . and a war-profiteer."[143] Designated a Class III activist, Patin claimed he joined the party in May 1933 "because it was the fashion among industrialists in Berlin at that time, and [I] always maintained only nominal membership in party." Patin attributed the honorary SA membership to having won the prestigious Lilienthal Prize in aeronautics in 1934. The army's security report highlighted Patin's value to Paperclip: "He has acted as leader and spokesman of other scientists and has rendered valuable aid in controlling the other scientists. He has been instrumental in maintaining high morale and keeping peace among the scientists."[144] The JIOA praised Patin for issuing nearly one hundred patents and applications related to navigation and remote control and, rejecting Nehlsen's opinion and other evidence, claimed, "Mr. Patin repeatedly risked his own position and reputation, even his own safety by refusing to cooperate with the Nazi party program." As an industrialist producing for the Nazi war machine, Patin employed what he termed "voluntary" foreign workers, including five hundred Jewish women, whom Patin supposedly aided to the best of his ability. Patin ordered that the "500 pitifully dressed women with shorn hair" be fed, provided medical attention, and given scarves for their heads. "I believe these people saw their first ray of hope when after a series of bitter disappointments they were employed in my organization," Patin related.[145] Patin was also arrested in June 1941 by the Gestapo, although Luftwaffe general Ernst Udet secured his release a day later. These instances combined with the air force's admiration for Patin's work mitigated the negative information in his record.

Herbert Axster

Herbert and Ilse Axster, like Rickhey, generated negative publicity for Paperclip.[146] The Axster saga, which cannot be treated exhaustively here, also underscores the problems spouses and dependents posed for US authorities processing immigration files as well as the Paperclip recruits trying to assimilate. Herbert Axster had served as chief of staff to Dornberger at Peenemünde, but he had been a patent attorney and Wehrmacht officer, not a scientist, and it appears his relationship with Dornberger and Wernher von Braun had more to do with his selection as a Paperclip prospect than his expertise in a relevant field. Despite Axster's being labeled a "notorious supporter and profiteer of the Nazi regime from 1933 to 1945" by OMGUS, the JIOA and related intelligence agencies worked to soften the assessment and secure a Paperclip contract in the United States.[147]

For all the derogatory entries in his personnel record, including denouncing Jewish lawyers and his possible involvement in burying documents and deceiving US authorities, Herbert Axster's greatest liability was his wife.[148] While Herbert Axster's presumptive classification for denazification proceedings was that of a "follower," investigative authorities recommended Ilse Axster "be classified as a Class II activist."[149] The picture of Ilse derived from affidavits and various accounts is a veritable caricature. OMGUS investigators named Ilse Axster a potential security threat for the following reasons:

A. She was an ardent believer and propagandist for National Socialism.
B. She has demonstrated her gullibility and readiness to ardently accept and promote new or radical ideas.
C. She was an arrogant and swashbuckling type.
D. She did not supervise and care for family workers in acceptable American standards.[150]

According to several eyewitnesses, Ilse Axster had managed the family's small farm outside Peenemünde and had regularly beat and otherwise abused foreign workers. Locals described her as "a rough sort of woman" who struck fear in the local population.[151] As the regime collapsed, she "summoned all the women of the town, held enthusiastic speeches about the merit and importance of National Socialism and gave instructions on shooting," encouraging them to "await the Russian troops and defend the Fatherland even with kitchen knives and spades," while she escaped with her husband.[152] Ilse responded to the charges, claiming to have protected Jewish children in a kindergarten and denying beating her workers, although she admitted to slapping an "impertinent" Ukrainian who refused to work.[153]

Ilse Axster's toxic reputation complicated Herbert Axster's career prospects, but she also served to shield her husband's own record from greater scrutiny. Herbert Axster's party record, the OMGUS security report read, "is regarded as insignificant in that it was probably due to his wife's influence and to the fact that most Germans of his financial position found it advantageous to be members. The only objection here to his immigration stems from the fact that he apparently condoned the actions of his wife."[154] Eventually Ilse Axster's antics became public, and no less a figure than Rabbi Stephen

Wise of the American Jewish Congress named the Axster couple in his condemnation of Project Paperclip.[155] The air force valued Herbert Axster, but retaining his services came at a price. Once again, the JIOA's screening process had left something to be desired at a critical moment in Project Paperclip's history. Only after a rift developed between Wernher von Braun and Herbert Axster did the military concede that the Axsters would have been better left in Germany.

Dreamers, war criminals, and, most commonly, technocrats composed the approximately fifteen hundred scientists and technical personnel acquired during Project Paperclip and its successor programs. The duration of the program is partially explained by the ease with which the Germans assimilated into the US MIC. The American perception of German scientists was hardly uniform, but most bureaucrats, military officials, and scientific organizations activated for total war absolved the Third Reich's scientific community of responsibility, even if other segments of American society did not. The US National Academies of Sciences described German science as "an island of nonconformity in the Nazified body politic," which withdrew into the "traditional ivory tower [that] offered the only possibility of security" in an oppressive regime.[156] Scientists who had retreated into an ivory tower did not interest American military authorities, and most Paperclippers had furthered their own dreams by realizing the Third Reich's. The view of German scientists as amoral technocrats hardly disqualified them from Paperclip. On the contrary, the more single minded and reliant on patronage the scientists appeared to be, the better they integrated into the US national security state.

CHAPTER TWO

Implements of Progress

The Military's Case for Paperclip

> Occupation of German scientific and industrial establishments has revealed the fact that we have been alarmingly backward in many fields of research. If we do not take this opportunity to seize the apparatus and the brains that developed it and put the combination back to work promptly, we will remain several years behind while we attempt to cover a field already exploited. Pride and face-saving have no place in national insurance.
>
> —Major General Hugh Knerr, June 1, 1945

In early 1945 intelligence and technical teams assigned to Anglo-American forces descended on the disintegrating Third Reich determined to locate and secure Germany's "wonder weapons" before they wreaked more havoc. Separately, the US Alsos mission scrambled to uncover the extent to which the Nazis had progressed toward producing their own atomic bomb. Samuel Goudsmit, the primary Alsos investigator, recalled the exigent circumstances of his mission: "We had obtained intelligence data on the V-1 and V-2. What final use could they be to the Germans unless they were meant to carry atomic explosives?"[1] Hitler's infamous March 1945 Nero Decree, pledging a horrific *Götterdämmerung* (catastrophic collapse) for invading forces was no empty threat, especially after months of indiscriminate V-1 and V-2 attacks on London and northern Europe during the war's endgame.[2] Neutralizing the missile sites, equipment, and personnel responsible for unleashing a last barrage of nihilistic destruction was the priority, but the war in the Pacific was still raging with no end in sight. Did Germany provide Japan with its revolutionary military technology, and if so, when? Could the Allies quickly and effectively harness Nazi technology to achieve a decisive victory against Japan? Addressing these intelligence gaps consumed operatives in Europe as well as Pentagon planners still engaged in active war fighting, but Major General Hugh Knerr articulated the strategic interest in exploiting the science and the scientists responsible for Nazi Germany's formidable war machine. Not everyone agreed that Nazi science was as advanced as Knerr assumed, but the bureaucratic consensus favored Knerr's position, including his admonition to the American scientific community about hampering the effort through "pride and face-saving." In Knerr's view, the infrastructure that had enabled the United States to become the world's most dominant military force in a few short years was apparently deficient. Nazi science, if properly employed, could secure US hegemony for generations to come.

In this chapter, I first describe the twisted road to forging a consensus around exploiting German science and scientists, a task complicated by conflicting military and civilian assessments of German capabilities produced during the waning months of World War II and the immediate aftermath. Initial investigations into German equipment and expertise were disorganized, haphazard, and plagued by bureaucratic confusion and rivalry. Second, I recount the deliberations authorizing Project Paperclip and the conflicting views on its scope. Third, I detail how the JIOA convinced the intelligence community and the national security bureaucracy to expedite the immigration of recruited scientists, despite inadequate or derogatory security investigations. Fourth, I examine why the United States chose not to pursue a Paperclip program for Japanese scientists. I conclude this chapter by addressing the British and French exploitation programs in relation to Paperclip.

Pillaging the Reich

During the course of planning Operation Overlord in summer 1944, General Dwight Eisenhower directed the G-2 (intelligence) branch to establish a T (technical) subdivision responsible for gathering intelligence on scientific and industrial targets, determining the extent of German-Japanese technology exchange, and locating and interrogating German scientists and technicians before they could dissolve into the population or find refuge abroad.[3] Eisenhower feared the effects of German missiles in particular and prioritized their elimination from the battlefield: "It seemed likely that, if the German had succeeded in perfecting and using these weapons six months earlier than he did," he wrote later, "our invasion of Europe would have proved exceedingly difficult, perhaps impossible."[4] The T-Forces accompanied combat units in the Sixth, Twelfth, and Twenty-First Army Groups as they penetrated German territory. T-Forces were military units, but the targets were selected by British and American civilian and military officials advising the CIOS, which produced an impressive 3,377 reports on targets in one year before disbanding in summer 1945.[5] One participant attributed the volume of reports to the German penchant for bureaucracy: "The German love of keeping things in a good order helped us immensely because they seemed loathe to make a mess and destroy what they were working on."[6] The more sites the T-Forces investigated, the greater their estimation of German advances in military technology.

Acquiring information on German-Japanese collaboration was Allied intelligence's top priority. The mission appeared more urgent after the German submarine U-234 was intercepted off the coast of Canada on May 14, 1945, on its way to deliver advanced technology to Japan. The submarine's cargo contained an alarming inventory of items: electronic torpedoes, a dissembled Me-262 jet aircraft, a Henschel Hs 239 glider bomb, and twelve hundred pounds of uranium oxide. Hidden among the crew were civilian engineers, scientists, and two Japanese officers trained in technical matters.[7] Although the uranium oxide prompted fears of a developed nuclear program in either Germany or Japan, the material was likely intended for aviation fuel.[8] In early 1943, Hitler granted the Japanese access to material and knowledge concerning Germany's operational weapons. In January 1945, as the Reich crumbled around him, Hitler agreed to share experimental technology as well. U-234 was Germany's last gift to Japan. Allied investigations into German archives revealed that the Japanese sought German equipment to bolster air defenses, specifically the Natter interceptor rocket, and arranged for Japanese engineers to train in Germany and German engineers to travel to Japan.[9] Yet, none of these plans materialized in time to make a difference.[10] One intelligence report claimed the Japanese sent a submarine to retrieve a German radar expert, although this too proved impossible given the advanced stage of the war.[11]

As the invasion of Germany transitioned into an occupation, Eisenhower sought to combine the disparate intelligence efforts relating to scientific exploitation and devote greater resources to safeguarding the seemingly endless list of names and locations. "The numerous military and non-military, economic, financial, scientific and industrial/technological activities in Supreme Headquarters Allied Expeditionary Force urgently require coordination, integration and firm direction," Eisenhower wrote. He lamented the "duplication of effort and loss of time" and refashioned the T-Forces into an "interservice integrated body" better known as FIAT.[12] FIAT was charged with the "conservation, exploitation and appropriate disposition of technical information concerning Germany" to achieve the "effective and lasting control of Germany's war potential." FIAT also sought to restore enough German industry to meet the basic needs of its people and provide reparations to the Allies. The "proper exploitation" of German science and technology may be the only "material reward of victory," one memo declared.[13] Eisenhower contended with multiple and sometimes conflicting missions related to the German scientific community. Germany must be disarmed, everyone agreed, permanently perhaps, but so too did it need to stand on its own as an industrial state in the heart of Europe. The process of determining when rebuilding begins and exploitation ends consumed military and civilian authorities, even after the formation of an independent German state.

FIAT had an inauspicious beginning, as teams arrived without uniforms or other identifying clothes, with no itineraries, and with inadequate transportation and accommodation. "They are unbriefed," a July 15, 1945, report read. "They have no access to the latest information (if it exists) on their particular subject; since either they have not been allowed access to the report on the subject, or they have not known where to go for the information."[14] This disorder ultimately empowered the detained Germans, as Walter Jessel noted in his screening report of the rocket team. FIAT eventually comprised 200 personnel, 123 military and 77 civilian. FIAT's specific missions included coordinating and directing technical investigations by US and other Allied personnel, locating, microfilming, and distributing technical manuscripts "of value to US science and industry," and supervising the "interrogation of scientific or industrial interest in US controlled areas of Germany."[15] FIAT's final report in June 1947 claimed that the organization "paid for itself in securing the allocation of war booty, such as rockets, war chemicals, aircraft and accessories, and wind tunnels."[16] General Lucius Clay, the military governor of the US zone, praised FIAT's two-year effort, particularly gathering and disseminating the technical data: "We took this information first to facilitate our war effort against Japan and then definitely for commercial purposes. This taking of this information to my mind . . . is parallel to Soviet action in taking current production and to French action in removing capital equipment apart from reparations. . . . I believe that the work accomplished by FIAT will prove to be mainly reparations to the US from Germany."[17] Project Paperclip advocates, many of whom were assigned to or worked with FIAT in occupied Germany, similarly touted the financial benefits of exploitation to silence critics wary of reconstituting German war potential.

The combined experience of the T-Forces and FIAT advanced the general impression that captured German personnel were more valuable than hardware. A member of the army air force LUSTY (Luftwaffe secret technology) exploitation team recalled, "[W]e had found out more about what had been going on in the war in a few days [*sic*] conversations with some of these key German leaders, than all the running around and digging for drawings and models . . . could bring."[18] In January 1945, SHAEF G-2 compiled a list of noted German scientists working on high-profile projects, an effort greatly assisted by Werner Osenberg's capture. Locating and securing a site was infinitely simpler than negotiating the complexities of separating "Nazis" and "war criminals" from similarly imprecise categories like "scientists" and "technicians."[19] FIAT eventually steered an effort to exempt scientists from denazification laws forbidding their continued research, recommending that "the term 'to utilize' rather than the term 'to employ' be used, since the mission of FIAT requires a greater range of liberty than is implied in the term 'to employ.'"[20] In other words, some scientists were to be interrogated briefly, usually in detention, and compensated with extra rations or housing. Others, such as the rocket team, were offered military contracts and the promise of eventual citizenship.

After months of collating data and deciphering blueprints and captured equipment, ranging from torpedoes to pristine jet engines, technical intelligence personnel communicated their opinion in a June 1945 progress report that "optimum results will be obtained when man, equipment and records bearing on a single problem are examined concurrently at the same place."[21] The military first considered a plan to assign enlisted men to "serve as apprentices" in various German enterprises like IG Farben, "for the purpose of learning techniques and processes, commonly known as 'know how.'" Some US civilians recommended the transfer of Germans to the United States in August 1945, the same time the army clandestinely smuggled the rocket team to Texas. Ironically, the civilians believed such a venture was "impractical for the present at least." The Industry Division of OMGUS, responding to reports of Soviet and French seizures of people and equipment inside their occupation zones, concluded that a "considered plan of absorbing German 'know how' . . . would offset criticism that 'we have captured everything except Germany's brains.'"[22] That considered plan evolved from working alongside Germans in their own labs to the once unthinkable prospect of shipping both individuals and equipment to the United States indefinitely.

The armed services and many businesses were clearly enamored with German science, but any sustained and meaningful use of that science would require maneuvering around the military's own constraints. The April 1945 directive delineating American occupation policy—JCS 1067—essentially called for deindustrializing Germany. The directive forbade all scientific research that could be used for military purposes, but the army violated its own directive almost immediately by hiring scientists engaged in Nazi-era military research projects to continue working on German soil. A year later the Allied Control Authority (ACA) reaffirmed the ban on research with Law 25, although enforcement was lax.[23] Certainly JCS 1067 had its critics, not the least of which were the primary officials responsible for its implementation. Lucius Clay's economic adviser, Lewis Douglas, said the directive was "assembled by economic idiots" who would "forbid the most skilled workers in Europe from producing as much as they can for a continent which is desperately short of everything."[24] Intelligence officials perhaps avoided breaking the letter of the law, but not the intent. "JCS 1067 restricts research but logically places no restriction on the personnel qualified to do this type of research since the same personnel are qualified to perform scientific duties clearly acceptable within the scope of the directive," reads one September 1945 FIAT memo. "By removing existing restrictions," it continues, "this category of personnel will be able to seek work in fields where they are particularly adaptable and the provisions of JCS 1067 can be enforced by policing the work rather than the abstract thinking of scientists."[25] Occupation authorities worried about feeding Germans and preparing for a future in which the defeated power could achieve self-sufficiency. Addressing these concerns required putting scientists back to work, but the armed services preferred that Germany's "best and brightest" remain experts in their wartime fields, not peaceful pursuits.

Admiring the accomplishments of German science during the Third Reich did not always imply support for reconstituting the scientific community, let alone implementing a policy as unorthodox as shipping enemy combatants to the United States to continue research outlawed in Germany. In June 1945, FIAT published a paper that specified the problems of reviving Germany's research capabilities and urged restraint. The report reminded readers of the irredeemable "perversion of German research" and the nation's last half century of harnessing "all the activities of men's minds to the problems of war." Referring to Germany, the report noted, "She was perhaps the only nation who carried prostitution of science to this extremity, . . . and these facts must be taken into account in any consideration of how best to deal with German research."[26] FIAT's recommendations focused on occupation procedures inside Germany, not the viability of Project Paperclip, but much of the report's content indirectly challenged the wisdom of the Paperclip scenario, which was already taking shape that summer. The report predicted that the future of warfare depended on weapons the Germans had pioneered, maintaining that allowing scientists free rein to innovate and experiment was ludicrous: "In fact, if one could be sure that German imagination would extend no further than projects for the improvement of existing weapons, for example, artillery, tanks, aircraft, rockets and the rest, it would certainly be a matter of rejoicing." Furthermore, the report rejected "the formation of coordinated networks of research and development organizations" and recommended that "the centralized direction of research should be prohibited."[27] Having identified the formula for Nazi achievement in advanced weaponry, FIAT reasoned the best course of action was to dismantle it piece by piece.

Paperclip depended on German scientists continuing the same research agenda for the US military that they had pursued for the Nazi state. FIAT, perhaps reacting to rumors of a Paperclip program on the horizon, concluded its June 1945 report with words of caution: "The possibility of harnessing German scientific and technological effort to the service of Allied industry should be carefully examined."[28] In September 1945, shortly before Paperclip's official authorization, FIAT informed the JIOA that "most of us believe great care must be taken not to let them [German scientists] carry on activities that would, if it were done in Germany, contribute in any way to Germany's war potential." Furthermore, FIAT asked, how can US counter-intelligence personnel determine whether scientists working for the United States remained loyal to "their 'fatherland'?" It was self-evident to FIAT that any proposal to hire and then repatriate German scientists was "contrary to the spirit of JCS 1067."[29] The directive presumably no longer applied, however, if scientists left Germany and became US citizens.

The US Foreign Economic Administration (FEA) published similar reports recommending strict controls on German science and technology. A multiagency initiative, the FEA established thirty-two disarmament committees to execute President Franklin Roosevelt's instruction of "seeing to it that Germany does not become a menace again to succeeding generations."[30] Several State Department officials involved in Paperclip worked for the FEA during the war and continued to privilege Roosevelt's wartime attitude toward German rearmament. In July 1945, while the army shipped the rocket team to the United States under the cover of night, the FEA issued a report critical of German scientists and technology related to "secret weapons." After detailing Germany's recent history of deception and secret rearmament after World War I, the FEA urged the complete dismantlement of the German MIC.[31] Charged with investigating secret weapons in the Third Reich, the FEA demystified German progress, attributing Nazi success to "Application Engineering and a surplus of 'trained minds and financial support.'" Using the V-2 as an example, the FEA identified the successful combination of industrial technicians, military technicians, and managerial technicians. "[M]anagerial direction is particularly necessary and, in general, stems from the military components of the central government."[32] While not named in the report, Wernher von Braun embodied the managerial technician elevated to "project manager." The FEA wrote admiringly of the German willingness to lavish pure research and place "scientifically strong people in positions of authority," although in Germany's case, such policies resulted in the militarization of the scientific profession.[33]

The FEA envisioned altering the structure, training, and culture of German science by eliminating military research completely. The first step toward accomplishing this, the committee argued, was eradicating the myth of the "great German inventive mind." The FEA clearly considered the US military as susceptible to the myth as German citizens and recommended that the government consider the risks of importing German scientists to the United States for any length of time. The FEA cautioned that German science could one day overcome its isolation and acquire valuable information from former enemies. Given Germany's experience evading the Treaty of Versailles after World War I, why tempt fate a second time? The FEA suggested that hiring German experts "will have the psychological effect of perpetuating the 'super race' idea in the minds of the German people."[34] Many scientists regarded employment with Allied forces as an opportunity to propagate this very image. Joseph Baur, a German engineer who assisted the US Army with the Me-262, argued, "The people of Germany had the right to know someday that during the darkest days of their existence something worthwhile was accomplished by their brilliant scientists and engineers."[35] The FEA clearly opposed any program like Paperclip, but it recognized some degree of exploitation was inevitable. The FEA concluded that any proposed use of scientists by the United Nations for military reasons should be carefully managed and short term: "They should be under military surveillance and guard adequate to prevent their acquiring knowledge of other projects and developments in this country [the United States] which have military uses, and should be returned to Germany upon completion of their specific assignments."[36] The FEA challenged the notion that German weapons experts had anything enduring to offer the United States. This skepticism continued inside the State Department during deliberations over Paperclip. Paperclip proponents inside the nascent national security bureaucracy, on the other hand, embraced the myth of the great German inventive mind and regarded security risks as either manageable or irrelevant.

Project Paperclip: Origins and Early Implementation

Project Paperclip coincided with the advent of the Cold War and the military's articulation of a national security doctrine, the core of which envisioned a permanent MIC supported by a substantial scientific infrastructure. The idea for such an expansive exploitation program originated with a series of uncoordinated efforts initiated by Anglo-American forces in the wake of T-Force discoveries during the final months of the war.[37] The navy planned for the first transfer of personnel to the United States in January 1945. Secretary of the navy James Forrestal floated the idea to secretary of state James Byrnes, noting that a transfer "would help to fill in some of the gaps in our research activities." Furthermore, Forrestal wrote, "it would be highly desirable to broaden the scope of this project, and make plans for exploiting German scientists in areas of interest to industry."[38] Similarly, the army air force (AAF) LUSTY mission and the Army Ordnance Department's interrogation of the rocket team prompted individual officers of varying ranks to contemplate prolonged exploitation stateside. Eisenhower's FIAT directive brought some order to the myriad collection efforts, but FIAT's mission was limited to acquiring and disseminating data to military and civilian customers, often with the aid of detained scientists and technicians. The proposal to transport German personnel to Allied territory and continue their wartime research under contract required executive approval. Paperclip was uncharted territory, however. "Everyone wanted a piece of the cake," one naval officer observed, "but nobody wanted to hold the platter."[39] The military necessity argument deteriorated with the Japanese surrender, although the Soviet threat was quick to fill the void. Ultimately, it took commercial and industry interests to convince the White House to authorize such an unorthodox program.

Though Paperclip evolved into an ecumenical program encompassing civilian and military interests, the military administered the program. The JIOA targeted scientists from "wish lists" compiled by the intelligence community and integrated them into primarily military institutions, although commercial representatives worked alongside the Germans almost immediately. The JIOA comprised a military director from the navy or army with the rank of colonel or captain, a civilian deputy director from a nonservice agency, and an advisory board reflecting the membership of the Joint Intelligence Committee, or JIC (i.e., members of the Departments of Treasury, Justice, and Commerce, the Office of War Mobilization and Reconversion, and the War Production Board).[40] The State Department was a constituent member of the JIC, but its status with the JIOA seemed unclear. The confusion was more a product of the personalities involved than disagreements over the chain of command. Clarence Lasby writes, "In both membership and function, the JIOA reflected the government's shift of emphasis to the acquisition of knowledge which would benefit American industry," and further claims it was overwhelmingly civilian in membership.[41] Lasby is incorrect. The JIOA was a military operation with limited civilian input other than responding to JIOA requests for expedited action. JIOA committee meetings usually included all members, but military personnel handled daily operations. Furthermore, though the JIOA charter designated a civilian deputy director, the position was consistently filled by military officers.[42] The small JIOA staff relied on military units in Europe to locate and transport scientists and on the overburdened CIC and OMGUS to handle investigations and screen recruits. The JIOA remained a small organization even after Paperclip and its successor programs acquired a greater profile among civilian and military leadership. Consequently, the JIOA chairman, an officer of relatively junior rank given the mission, exercised an inordinate influence over a program with far-reaching consequences.

Project Paperclip was an unlikely prospect under a Roosevelt presidency. In December 1944, OSS chief William Donavan approached the president with a plan to utilize German intelligence personnel in the postwar period. Roosevelt responded negatively, noting that guarantees of protection and immunity from prosecution "would be difficult and probably be widely misunderstood both in this country and abroad. We may expect that the number of Germans who are anxious to save their skins and property will rapidly increase." Among these, Roosevelt assumed, "may be some who should properly be tried for war crimes or at least arrested for active participation in Nazi activities." Even with safeguards and screening, Roosevelt concluded, "I am not prepared to authorize the giving of guarantees."[43] Roosevelt harbored some of the same suspicions about ex-Nazis as did the perceptive Walter Jessel. Furthermore, Roosevelt did not oversee the deteriorating relationship between the United States and the Soviet Union, nor was he likely to jeopardize domestic security by inviting hundreds of enemy aliens to work in the defense establishment.

Similarly, undersecretary of war Robert Patterson, later a staunch advocate for Paperclip, first expressed his doubts to chief of staff Admiral William Leahy. Patterson supported using German scientists in the war against Japan, but cautioned, "These men are enemies and must be assumed they are capable of sabotaging our war effort. Bringing them to this country raises delicate questions, including the strong resentment of the American public, who might misunderstand the purpose of bringing them here and the treatment accorded them."[44] Once apprised of the program's potential, however, Patterson overcame his objections and promoted Paperclip throughout the bureaucracy. The War Department expected Americans to undergo a similar conversion once Paperclip's benefits were advertised to a skeptical public.

President Harry S. Truman approved the first exploitation—Operation Overcast—in July 1945. After numerous security breaches, the military changed the name to Paperclip in November 1945. Operation Overcast authorized the transfer of more than 350 rocket scientists to Fort Bliss, Texas. The civilian scientific adviser to FIAT wrote excitedly about using the detained scientists: "The exploitation in the US of the technical group from Peenemünde would seem to be a project which has been decided upon in higher quarters, presumably on the grounds that this group has technical information and abilities which can be used to further weapons development in the States for use against the Japs!"[45] Colonel Holger Toftoy of the Army Ordnance Department understood military necessity was paramount, but once the Japanese threat was extinguished, he touted the combined scientific and military value of Project Hermes, a plan to construct and test one hundred V-2s at the newly constructed White Sands Proving Ground in New Mexico.[46] The US effort coincided with a joint Anglo-American initiative, Project Backfire, which gathered hundreds of V-2 experts and technicians to assemble and launch several surviving V-2s near Cuxhaven in Lower Saxony.[47] Anglo-American cooperation continued for the most part, but the overlap between Overcast and Backfire led to a dispute over eighty-five scientists targeted by both countries, sparking the first of many incidents prompting greater coordination between the two Allies.[48]

Overcast required a unified effort from military and civilian agencies to ship hundreds of scientists and technicians, along with boatloads of unassembled V-2s, across the ocean, transport both across the country, and invite relevant private industry experts (General Electric, for example) to maximize the value of the project. Smaller, more secretive programs progressed nearly undetected. Project 77, initiated by the Special Devices Division of the navy's

Office of Research and Inventions, used German scientists to assist with captured technology. In July 1945, Captain W. B. Phillips informed the FBI field office in New York that the project was rated "highest secret" and included just five scientists "who are being brought to New York to work on a secret weapon which is presumably to be used against the Japanese."[49] Project 77, which was housed in a secluded Long Island estate, eventually increased its personnel to sixty scientists. An internal FBI memo informed J. Edgar Hoover of the unusual project on American soil: "These people are supplemented with German prisoners of war who because of their background might assist in this technical undertaking. Shortly there will be added to this German group a similar number of Japanese for like purposes." Thirty Japanese scientists were expected to join, but it is unclear whether they ever arrived.[50] The FBI had no role in the project, but the navy informed the agency that it intended to allow scientists to leave the property "for entertainment purposes."[51] Details of Project 77 are scarce, but the "secret weapons" likely related to the capture of U-234 and its German-Japanese cargo. Heinz Schlicke, an electronics and infrared expert, was on board the submarine and counted among the Project 77 scientists. Enemy alien civilians were required to return

V-2 rocket tail fins in transport, Allied-occupied Germany, June 1945. The disassembled rockets would ultimately join German rocket scientists in the United States. *Source:* United States Holocaust Memorial Museum, courtesy of James Baker, 01275.

to Germany in July 1946, but the navy was informed in April 1946 of a government program intended to "provide for the future employment of German scientists and technical personnel" after the contract ended.[52] The program was part of Project Paperclip and it ensured that programs like Project 77 would continue uninterrupted.

The scope of the original Paperclip program expanded in March 1946 to include the possibility of citizenship and long-term contracts for scientists if the sponsoring agencies deemed such an offer "in the interest of national security." Immigration extended to scientists' dependents as well. The March directive included reduced security measures, now that hostilities had concluded, and released the Paperclippers for consultation and eventually employment with civilian industries and laboratories. Alarmed by reports of Soviet recruitment of German scientists and under pressure from private industry and secretary of commerce Henry Wallace, Truman agreed to raise the number of scientists allowed from 350 to 1,000 in September 1946.[53] As early as June 1945, David Sarnoff of the Radio Corporation of America urged Truman to bring over as many Germans as he could before the Russians did.[54] Truman was initially reluctant to approve a postwar industrial and commercial exploitation program, telling Vannevar Bush he "was morally certain that our home boys would not want any competition," but the pressing need to deny the Soviets access to scientists—the national security argument—won the day.[55] Truman and others suspicious of the program eventually accepted a broader view of national security and welcomed the commercial benefits of integrating one thousand German scientists.

As powerful as the denial argument based on Soviet threat was in some circles, the prospect of building a MIC equal to the nation's global responsibilities resonated more with senior civilian leadership. Henry Wallace, the great progressive and Cold War opponent, nonetheless supported a vigorous Project Paperclip and lobbied Truman in his role as secretary of commerce. Wallace fielded multiple requests from corporations anxious to access German science, including Bell Telephone Labs, Rangertone, Western Electric, Dow Chemical, the American Optical Company, the American Chemical Society, and the Scientific Apparatus Makers of America, to name a few.[56] Bradley Dewey of the Dewey and Almy Chemical Company praised FIAT's microfilm effort but requested more. "If American industry is to copy much of the work of the Germans," Dewey wrote, "it will want to re-engineer many of the processes to fit our industrial pattern. . . . It is for this reason that it is so imperative that we have access to original lab reports, pilot plant data, design calculations, engineering calculations, economic studies, drawings."[57] Dewey claimed, "The boost to our economy and national defense of just one or two of the ideas that have been worked out in Germany will pay many times over the cost of the entire investigation."[58] Wallace wrote Truman an influential memo, "Proposed Importation of German Scientists for US Science and Industry Benefit," in late November 1945 in which he described Paperclip scientists as "intellectual reparations" and "the most practical and enduring national asset we can obtain from the prostrate German nation." Wallace's conception of Paperclip envisioned the "transfer of outstanding German scientists . . . for the advancement of our science and industry." The number of "eminent scientists whose contributions, if added to our own, would advance the frontiers of scientific knowledge for national benefit" numbered, in Wallace's estimation, no more than fifty. Wallace also expected a "careful screening" and open access to whatever knowledge they possessed.[59] The military, however, envisioned a thousand scientists and adhered to its own standards for what constituted "careful screening." Truman approved Paperclip just days later.

American intelligence personnel ranked targeted German scientists according to their value to US military and industrial needs in the same way OMGUS classified Germans citizens in its denazification program. The categories devised by Paperclip administrators indicate they looked beyond military necessity from the outset. Category I comprised those who "demonstrate unique talents" for the War Department; Category II scientists were "useful to industry" but indirectly to the military; Category III were "outstanding in their fields" and useful to both the military and industry; Category IV included those "not confirmed of value," but who had relevant credentials.[60] After getting approval for 350 scientists, the War Department began selling the State Department on the idea of transferring up to 1,000 to the United States for an indeterminate period of time. "[T]he State Department should be acquainted with the problem involved in exploiting the technicians," one officer noted. "If any large scale exploitation is contemplated for industry the State Department must necessarily adopt a stand."[61] On December 13, 1945, acting secretary of war Patterson sent a letter to secretary of state James Byrnes using the language of national security: "In those areas where German science, technology and industrial techniques are conspicuously in advance of our own or unique there rests upon the US the obligation to transplant these implements of progress, fitting them into our own scientific, technological and industrial structure." Patterson needed the State Department to facilitate the most controversial aspect of Paperclip—immigration and citizenship. "It is felt that the benefits to be gained by our nation would greatly outweigh any danger," Patterson continued.[62] A shorter message between War and State noted the primary objective of Paperclip "is improvement of our weapons," but "[i]t would be a very definite loss to our industry and indirectly to our future war potential not to exploit outstanding German scientists and technicians in the broad industrial field."[63] Paperclip proponents believed the program was a national asset requiring State Department investiture and partnerships with the private sector, not a narrowly conceived military program operating in the shadows.

The air force and the aviation industry modeled the symbiotic relationship between government and commercial enterprises characteristic of the MIC in the early postwar period. Senator Albert Thomas argued, "We won the war with the Germans with brawn not brains. We choked them with the weight of our planes."[64] In the future, many argued, we needed both. First on the ground in the defeated Reich, AAF officers pressured civilian and military bureaucracy to open the floodgates for German scientists and technicians. A German engineer recounted his experience with recruitment, noting his first contact with an American in June 1945 was a "Mr. Kinneman from Boeing aircraft company" looking for Junkers aircraft personnel. The engineer followed Kinneman to an office to meet with a second lieutenant and a CIC agent: "They said the American team was interested in my expert knowledge and wanted to know if I would work for them. My immediate family and my close associates could come along. I could take the most important things (documents and equipment) for my work, but there was no need to bring extra clothing and other personal things."[65] The Junkers engineers, like other talented and displaced German scientists, provided the AAF affordable and plentiful labor, especially since so many American scientists working for the air force returned to the private sector after the war. Moreover, the air force believed German civilian scientists were easier to control than their American counterparts.[66] Colonel Donald Putt, who oversaw Paperclippers at Wright Field in Ohio, noted, "Many of the German scientists have from six to ten years of employing techniques and equipment not yet available in this country." Putt acknowledged that the United States might build beautiful research labs, but "America is definitely short-handed in the other requirement for outstanding development—BRAINS."[67] The military assumed that scientists of every nationality existed in a state of "technocratic innocence," defined by Mitchell Ash as "the belief that science and technology are essentially apolitical, value neutral instruments or tools, easily transferrable cultural capital."[68]

The perception of Germans' "technocratic innocence" resonated throughout the national security bureaucracy and justified whatever shortcuts necessary to realize Paperclip.

Representatives from the Aircraft Industries Association of America, the Curtis-Wright Corporation, and the West Coast Aircraft Industry contacted AAF Headquarters and Wright Field to interview Germans on-site and, more importantly, direct Paperclip recruitment in Germany. One industry letter to General Henry "Hap" Arnold, the commanding general of the AAF and guiding force behind Paperclip, communicated the "urgency of securing the services of these men in the US at the earliest possible date," even if for a short consultation. Industry leaders made it clear, however, "that permanent or semi-permanent employment in the industry would be still more effective in the most rapid development of advanced aircraft and guided missiles."[69] The navy vied for its own specialists, noting a "nucleus of creative armament designers . . . will enable the gap between armament and airplane to be closed."[70] The Commerce Department agreed, stating the "intelligent exploitation" of German aeronautical experts "should prove highly beneficial to the American aviation industry. This group is a source of a specific type of information which is not likely to be found available elsewhere."[71] The AAF lobbied successfully for a powerful, independent, and fully funded air force to defend the nation from "wonder weapons" and project power globally. General Arnold predicted the future of atomic warfare belonged to the V-2, which he considered "ideally suited to deliver atomic explosives," and foresaw an era in which atomic bombs would be launched "from true space ships, capable of operating outside the earth's atmosphere."[72] Michael Sherry notes officers like Arnold were "shrewd propagandists, skillful at invoking all the terrifying developments of the war to dramatize their case" for funding and a privileged role in the national security state.[73] Project Paperclip helped the AAF realize this position better than any other program.

The JIOA Takes Control

The military soon discovered that selling Paperclip to the executive branch proved simpler than negotiating the labyrinthine national security bureaucracies. The JIOA was a powerful organization, excessively so given its ability to circumvent established protocols and a panoply of laws, but it necessarily relied on other agencies to implement Paperclip on both sides of the Atlantic. The JIOA's primary responsibility concerned administering security evaluations for the hundreds of scientists and dependents authorized to work in the United States. Executive authorization paved the way for extended employment, including the possibility of immigration, but strict standards remained in place for most of Paperclip's official acquisition phase between November 1945 and September 1947. JIOA officers worked diligently to ensure investigations of problematic cases included as much mitigating information as possible, but they preferred to omit any offending details. Party affiliations, war crimes accusations, and ideological fervor normally disqualified recruits from any extended employment inside the United States, let alone citizenship, but the JIOA arranged for expedited investigations that either ignored derogatory data or included amended security dossiers. Both military necessity and national security arguments resonated with bureaucracies attuned to the oncoming Cold War.

Colonel Howard M. McCoy led the AAF's LUSTY mission to investigate Luftwaffe technology and played an important role in administering Paperclip as chief of intelligence for the Air Material Command (AMC) at Wright Field. McCoy summarized his experience with the nearly two hundred German scientists assigned to AMC in a May 1947 memo intended to stave off funding cuts. McCoy divided Paperclip scientists into two types, "one the superlative specialist in his field, and on the other an excellent basic engineer. The specialist, in certain fields, is the best available in the world today. . . . Not to make use of this experience would necessitate considerable extra work and expense on the part of the US." McCoy believed Germany's basic engineers were "approximately equal in ability to the best US engineers" and justified their presence as supplementing losses to the private sector. "They would in no case compete with or take jobs from the War Department civilian employees."[74] McCoy then deployed the most effective weapon in the pro-Paperclip arsenal—denying scientists to the Soviet Union. Returning "a few" scientists to Germany may be inconsequential, McCoy wrote, but "the return of a great many, each of whom may have only a bit of retained knowledge, may result in the piecing together of all their information and disclose some of our vital secrets." McCoy cited the stream of reports concerning "Russian offers of jobs" to scientists returning to Germany. "It is believed very important to continue to deny to other countries the benefits obtained from the scientists now employed under Project Paperclip."[75] The military invoked the Soviet threat to rationalize the "expanded Interim Paperclip Program" authorizing 1,000 scientists compared to just 350 in Operation Overcast.[76] Denial justified shortcuts in the screening process, hiring mediocre engineers of questionable value, and violating US immigration laws.

General Hugh Knerr encouraged treating Paperclippers respectfully from the moment they were captured, writing in May 1945 that "we would get the most out of them by making them comfortable" and floated the idea of "sending their wives with them."[77] Knerr recommended paying Germans a "good salary" and rejected detaining scientists as prisoners or worse, "slave workers." "The scientific mind simply does not produce under duress," Knerr wrote.[78] Despite Knerr's preferences, working conditions at Wright Field were initially substandard in many ways. Germans lived in poorly constructed housing while their families remained isolated in the Landshut camp in Germany, although they still lived better than most Germans. At Wright Field, Germans complained bitterly, noting that the United States was not the only power in need of their services. Knerr and his subordinates were alarmed. Colonel Donald Putt, who managed daily operations relating to Paperclip at Wright Field, predicted the program "will be doomed to failure" unless conditions changed. Putt admitted, "[T]he thing has been badly managed and since we are competing with the British and Russians for the voluntary services of these people, we are now in a very poor bargaining position."[79] By fall 1945, after Paperclip's authorization but before immigration was viable, Putt expressed his frustrations to Knerr, blaming those "responsible for the establishment of existing policies" for an inability to appreciate the special problems associated with handling the scientists. Putt accused them of adopting "a prisoner of war attitude towards these German personnel." Putt, whose signature is on every positive security evaluation of scientists assigned to Wright Field, blamed resentful bureaucrats for mistreating his Germans: "Although it is difficult for the American mind to have much sympathy for the personal problems or difficulties of the German personnel, nevertheless cognizance of them is necessary in the treatment of Germans if we are to realize the fruit of their brains."[80] Officers like Putt and Knerr maintained that conferring citizenship to their Germans rewarded scientists properly and achieved the objective of denial to "foreign powers."

Colonel Putt resented Americans treating incoming Germans as POWs, but the staff at Wright Field supplied the Paperclippers with as many as 150 of their imprisoned countrymen to assist with projects and help out with "housekeeping" in their residences. Putt screened prisoners for special skills, much like the rocket team did with foreign laborers at Mittelwerk and Peenemünde, and resisted the "infiltration of qualified American personnel" into the Paperclip projects at Wright Field, at least in the short term. "To obtain the best possible contribution from the German scientists in the minimum of time," Putt wrote, "it is believed essential that an adequate staff of German engineers, technical assistants and draftsmen be maintained." American officers connected to Paperclip generally agreed that the Germans produced better results if conditions resembled the scientists' experience during the Third Reich. Putt expressed tremendous deference to the "great German inventive mind" so derided by critics of the exploitation program: "While scientific people are probably less prejudiced than many others, and are more adaptable to changing conditions, it is unlikely that even the best of them could be molded into a staff of assistants which could approach the smoothness of operation of a staff of German who have been trained from childhood in the German method." Therefore, Putt concluded, adequate prisoner labor was "absolutely essential to the successful operation" of AAF projects, especially since more Germans were expected to arrive as Paperclip progressed.[81] Experience conditioned American officers to invest Paperclippers with valuable resources and a remarkable degree of autonomy. The more the Germans produced, or at least promised to deliver in time, the more the military endorsed a permanent German presence in the national security infrastructure.

Legitimate security concerns unrelated to appeasing scientists prompted calls to transfer scientists' dependents to the United States. The camp at Landshut provided better food and conditions to family members compared to other Germans, but the facility was vulnerable to foreign intelligence surveillance, and not just from Soviet agents. The true purpose of the camp, housing Paperclip personnel and their families, was an open secret, and apparently so too were specifics of Paperclip. An April 1946 CIC report confirmed that Landshut was a sieve when it came to security. "The fact that so much is known about an operation that is classified as secret is a serious matter," the agent wrote.[82] That the most sensitive facts of the program were widely known inside the camp was an intractable security dilemma. The CIC surveyed "a widely scattered group of civilians in Landshut," all of whom "were acquainted with the general nature of the project." If this cross-section of the local population knew about Paperclip, one had to assume the surrounding towns were compromised. "That this situation exists is perfectly logical since the German civilians in the project have of necessity been allowed to go and come as they please, and to live in some sections of the city interspersed with other Germans."[83] Authorities expressed concern over "subversive elements" inside the camp as well, specifically among some of the women. "Two factions have arisen among the women of the camp," stated one inspection report. One "trouble maker" in particular provoked the natural tendency of women to form "feminine cliques." "[I]t is felt that the presence of such facts, if not actually or potentially subversive, will lead to unrest among the Germans at the camp and therefore should be closely watched."[84] Considering the untenable situation at Landshut, at least in the opinion of US intelligence services, transporting dependents and expediting background investigations of scientists awaiting contracts both removed the espionage threat and improved German morale.

Several hundred Paperclippers traveled to the United States in summer 1945 under military custody and armed with short-term contracts. Given Paperclip's potential to improve research and development in the United States and considering the worsening relations with the Soviet Union in Europe, the JIOA prioritized immigration above all else. The Paperclippers already working in the United States for six months, the length of the original contract, had priority, but the JIOA expected hundreds more to sign contracts after September 1946. The JIOA expedited the process by sending an officer to Wright Field and other locations with a concentration of Paperclippers "to assist the Air Corps in the preparation of papers to constitute dossiers for the first group of scientists to be recommended [for immigration]."[85] This assistance constituted evading normal denazification procedures and contextualizing the inevitable negative background information in the security dossiers. The JIOA was sympathetic to the plight of German scientists compromised by Nazi Party affiliations or worse, and emphasized the negative national security implications of following the letter of existing US laws. "The fact that both the Russian and French governments ignore the German Scientists' previous political affiliations has a demoralizing effect upon those whom the US consider possible candidates for exploitation," the JIOA explained. "Those in the American zone have knowledge, through letters and correspondence from friends, of ardent Nazis being comfortably settled in the Russian and French zones with their families, and continuing their research while those in the American Zone whom are considered for our Exploitation Program are stalemated."[86] The JIOA consistently invoked denial in its war with State Department officials over the immigration process, a topic explored further in chapter 3, but JIOA officers and empathetic liaison officers in Germany skillfully evaded what the JIOA considered a needlessly cumbersome security process.

The JIOA appeared to be at the mercy of OMGUS bureaucracy and reconstituted German governments empowered to conduct their own denazification regime, but intelligence priorities took precedence over "the German local government," according to a November 1945 FIAT memo.[87] The JIOA urged the War Department to acquire one thousand blank denazification questionnaires (Meldebogen) so that Paperclippers or dependents over the age of eighteen could complete the paperwork and be tried in absentia while continuing to work in the United States. This scheme required OMGUS convene a special tribunal.[88] The JIOA assumed correctly that it could control events better in the United States, but military governor Lucius Clay contested the need for any tribunal, calling the proposal for special tribunals "unwise and unfitting to jeopardize the operation of so important a program as Paperclip by subjecting it even partially to the whims and prejudices of German denazification agencies who might be tempted to obstruct or sabotage the program through delaying tactics or distortion of the facts." Clay also feared the consequences of giving scientists and technicians "special treatment" when other Germans had to endure the complete process. "It would be much better to permit them to remain in the US as Nazis without bringing them to trial than to establish special procedures not now within the purview of German law."[89] Clay's statement is a remarkable admission given his initial skepticism of a program that deprived Germany of its most talented minds during reconstruction. Clay's staff study foreshadowed tension between the United States and West Germany over the exploitation program, which continued under different names well into the 1950s. Unfortunately, the JIOA could not avoid the denazification process since a certificate clearing scientists was a prerequisite for immigration.[90] The OMGUS security report formed the basis for both denazification and immigration, but the JIOA had ways of influencing the content and conclusions contained in the reports.

The JIOA's goal of transporting at least one thousand scientists and technicians across the Atlantic was partially aided by the overtaxed OMGUS bureaucracy and undermanned military intelligence infrastructure. The CIC was particularly inundated with multiple missions, ranging from denazification, war crimes inquiries, and fulfilling its charter mission of counter-intelligence, to investigating scientists on the black list. After the OSS disbanded in October 1945, the CIC was left severely understaffed and unprepared for its myriad responsibilities. Consequently, Gerald Steinacher argues, CIC operations "often seemed amateurish and slightly desperate."[91] In January 1946, the War Department designated the Soviet Union and its military forces in occupied Europe the primary target for intelligence collection. In response, the CIC reassigned resources from hunting war criminals, denazification, and processing Paperclip security evaluations to the sprawling Soviet target.[92] The CIC felt particularly burdened by the Paperclip requirement, complaining to EUCOM that "were this organization to conduct the security investigations the administrative detail alone . . . would tie-up the bulk of the agents in the field."[93] The JIOA relied on the CIC for data collection, but neither agency was especially invested in discerning the truth behind any Paperclip recruit's personal history.

The CIC's procedure for vetting its own informants resembled the screening process for Paperclip recruits. The CIC agent first acquired a personal history statement, which usually followed the denazification questionnaire. Second, the agent verified information with local police, other agencies, and most importantly, the Berlin Documentation Center (BDC), which housed surviving Nazi Party and SS personnel records. The BDC was obviously incomplete, and records contained in the Soviet zone were either unavailable or censored. The JIOA often used this excuse when submitting incomplete dossiers to immigration authorities. The CIC also consulted the Central Registry of War Criminals and Security Suspects (CROWCRASS), described by Thomas Boghardt as an "unwieldy monster-archive containing over 100,000 names, filled with uncorroborated evidence and hearsay." CROWCRASS proved useless for expeditious background checks, but it satisfied the JIOA's low threshold.[94] While the international military tribunal at Nuremberg declared the SS, Gestapo, and other Nazi agencies illegal, the US intelligence community reserved the right to employ compromised Germans.[95] SS membership in particular usually resulted in immediate disqualification, but the JIOA exonerated affected candidates with creative contextualization and expected the rest of the bureaucracy to accept its judgment.

Like the CIC, the JIOA determined the denazification program too unwieldy and altered its own criteria from prohibiting "known or alleged war criminals" and "active Nazis" to the nebulous category of persons who might "plan for the resurgence of German military potential."[96] Truman's executive order for Paperclip forbade "active supporters" of National Socialism, but the JIOA defined the term to suit its own interests.[97] The JIOA hoped to prevent embarrassing revelations like those of the Axster couple and Georg Rickhey's indictment by coaching both scientists and army investigators on submitting "clean" dossiers. JIOA director Colonel Thomas Ford distributed a December 1946 memo informing staffers to first "indicate the relationship of the scientists to underground resistance or resurgence movements, and the degree of his participation in Nazi political organizations." Almost every applicant joined at least one party organization, but "the specific nature and ramification of the organization involved should be determined to insure [*sic*] that it is not the National Socialist or Pan-German category" outlined in denazification laws.

Most importantly, Ford wrote, the dossier "should include a statement that there is no probability that the scientist will be charged by any of the Allies as a war criminal, or be required as a witness to war criminal acts."[98] Field investigators were predisposed to clear the most desirable candidates and resented legal and bureaucratic obstacles to their work. Investigators working on behalf of FIAT and JIOA shared the opinion that most Paperclippers were forced to work under considerable duress in Nazi Germany: "Because we fortunately lack the concept of the Police State, we lose sight of the fact that there are very few men who will stop an express train by jumping in front of the engine. We'd think such a man a fool in the US."[99] In other words, German scientists' presumed "technocratic innocence" excused all but the most egregious acts of collaboration and criminality.

The saga with the State Department conditioned the JIOA to manage the security processes of "sponsoring agencies" and to neutralize likely objections. The JIOA once returned a stack of dossiers from Wright Field "for completion in greater detail. Most of the reports failed to indicate that any interrogation had been made."[100] A significant problem involved claims by several scientists that their membership in the Nazi Party, including the SA and SS, was "automatic" or "other than at [their] own request."[101] In the case of Ernst Stuhlinger, a senior member of the rocket team, the JIOA discovered that Stuhlinger had applied to join the party in October 1937, but in the JIOA dossier, he claimed that he'd had no choice in the matter. "Published facts about the membership procedure for the Nazi party indicate that every individual who became a party member had to fill in and personally affix his signature to the application form," wrote Kurt Lossbom of the JIOA's Exploitation Division. Lossbom requested that Stuhlinger submit another sworn statement "setting forth in detail the facts pertaining to his entry into the NSDAP either denying or acknowledging membership in the SA."[102] The JIOA was troubled by Stuhlinger's apparent dishonesty, not the false record, which could be explained or, in some cases, modified. Navy scientist Hermann Kurzweg's SS membership was left unexplained in his file, requiring the JIOA to once again "obtain more complete information on his past activities so that his application for Visa could receive full consideration."[103] Even Wernher von Braun's immigration visa was jeopardized without "a detailed report, including any extenuating circumstances involved in his membership in the SS and NSDAP. Complete information will be necessary since von Braun will undoubtedly remain in the US indefinitely to work on the rocket research projects."[104] For those scientists who admitted lying about their

Nazi connections, the State Department conceded "the mere admission by the alien that he committed under perjury . . . cannot be considered as perjury under the immigration laws and regulations."[105] The JIOA helped scientists draft convincing "extenuating circumstances" and evade civilian bureaucrats' objections, but some cases required more than revising paperwork.

In April 1947, senior State Department official Herbert Cummings supposedly "hit the roof" after learning of discrepancies between OMGUS security reports and the cleaner, more politically correct reports compiled by the JIOA.[106] The unfortunate JIOA staffer who met with Cummings reported "[t]hat data contained in the OMGUS security report is in most cases very sketchy and frequently does not agree with data contained in the biographical and professional form and security report by the sponsoring agencies." Cummings "inferred that the OMGUS reports are not complete" and "intimated that scientists who are members of the Nazi parties or its organizations prior to 1933 should not be considered for immigration."[107] Linda Hunt first discovered the JIOA's deceptive practice of submitting "revised security reports" intended to circumvent denazification courts and immigration authorities. "If State would not approve immigration due to derogatory OMGUS reports, the JIOA would change the reports to expunge derogatory information, in direct contravention of the president's policy," she writes. "Furthermore, JIOA officers knew they could rely on their counterparts in Germany to help them do the job."[108] Former Justice Department investigator Martin Mendelson told Hunt in 1985 that JIOA personnel concealed, or at least "failed to disclose," SS membership in reports to his office and to the State Department's Visa Division.[109] JIOA deputy director Walter Rozamus recalled in 1987 that he had never considered "changing a sentence or two" in the OMGUS reports deceptive. Rozamus also denied ever seeing Truman's directive declaring ex-Nazis inadmissible.[110] Linda Hunt presented evidence of revised reports in which the phrase "ardent Nazi" was changed to "not ardent Nazi" in an interview with JIOA officer Montee Cone. "So what?," Cone replied, claiming all he did "was pass the papers around." Although the interview occurred forty years after his JIOA assignment, Cone's explanation, or lack thereof, spoke volumes about how little the JIOA cared about a particular scientist's background. Cone told Hunt he was sure "people on the JIOA were anxious not to make a horrible mistake, but from a military point of view these people were invaluable to us."[111] Cone claimed the JIOA avoided errors and discrepancies, but intimated that even if they were purposeful, perhaps the reasons mattered more than obsolete laws. The cases of Kurt Debus, Emil Salmon, and Ernst Eckert demonstrate how far the JIOA went to secure a clean record for prominent Paperclip recruits.

Kurt Debus

Kurt Debus, an original member of the rocket team and the first director of the Kennedy Space Center, joined the SS in 1939 and served as a witness in proceedings against a fellow engineer, Richard Craemer, in 1942 and 1943. Debus denounced Craemer to the Gestapo for uttering anti-Nazi statements, and his testimony could have landed Craemer in prison for two years. Like many on the rocket team, Debus avoided a denazification proceeding in which he was expected to be designated Class III (minor offender) but possibly Class IV (follower) on appeal.[112] The JIOA contended with both the denunciation issue and the fact that Debus had lied on two separate occasions about his SS membership. The JIOA initiated a "reinvestigation" of the Craemer-Debus controversy in hopes it would have the same effect as the revised security reports in deflecting State Department attention. The JIOA learned from another Paperclipper involved in the Craemer trial that Debus's denunciation and testimony were coerced, and that Craemer actually apologized to Debus for assuming it was deliberate.[113] The second problem of SS membership was complicated by Debus making "no mention of SS rank" in his original JIOA form and later claiming he held a junior rank in a March 1949 sworn statement. In his 1947 denazification questionnaire, Debus listed a different rank, and the OMGUS report includes yet another.[114] The JIOA expended considerable resources on the reinvestigation and secured another personal statement from Debus explaining why he was confused over his SS rank. The JIOA felt "justified to issue a new Revised Security report," but conceded that the label "ardent Nazi" had to remain.[115] Immigration seemed improbable, but Army Ordnance and the JIOA invoked denial once again: "The alternative of returning Dr. Debus and his family to Germany involves considerations of national security, interest of the government and the specialist's demonstrated ability."[116] Moreover, denying talented scientists to foreign powers seemed even more urgent in 1950 than it had at Paperclip's conception in 1945.

Emil Salmon

Emil Salmon, a jet engineer assigned to Wright Field, presented the JIOA with a formidable problem. In July 1947, just a month after signing a Paperclip contract, Salmon was convicted of participating in the destruction of a synagogue in Ludwigshafen during Kristallnacht (November 9–10, 1938). The sentence was six months' hard labor and a Class II (offender) designation.[117] The JIOA and AMC scrambled to rectify the situation and at least delay Salmon's return until he completed his project. JIOA chairman Bosquet Wev recommended Salmon's case "should not be forwarded to the Department of State and Justice without a statement in the dossier that indicates the Air Force is cognizant of his Nazi activities and desires his immigration in spite of this knowledge."[118] The AMC pressured the JIOA to reopen Salmon's case before the denazification report "to secure a new sentence sufficiently mitigated to allow further processing of his case for immigration" and, remarkably, change his status from a Class II offender to a Class IV follower.[119] Salmon belonged to numerous Nazi Party organizations, held rank in the SA from 1933 to 1945, and was ideologically committed to the Nazi Party, as several associates verified.[120] JIOA investigators revisited the synagogue burning in May 1948 and determined Salmon's participation was unclear, despite the weight of evidence.[121] In June 1949 an OMGUS investigator delivered JIOA the decision it sought: "It is to be concluded that the accusation of Salmon is unfounded."[122] Reinterpreting the case evidence did not reverse the decision, but it allowed the JIOA and AMC to proceed as they saw fit. The AMC stated it was "cognizant of Mr. Salmon's Nazi activities and certain allegations made by some of his associates in Europe, but desires his immigration in spite of this."[123] Salmon's value to the air force superseded the damning paper trail detailing his past.

Ernst Eckert

The Czech-born Ernst Eckert researched jet propulsion at Wright Field before joining the faculty at the University of Minnesota in 1951. Eckert admitted to joining the SS in 1938 after first becoming a member of the pro-Nazi Sudetendeutsche Partei in 1937, but the real obstacle to his immigration status were accusations that Eckert had conducted cold-resistance tests on human subjects while working at the German Institute of Technology in Prague.[124] The JIOA learned of an incriminating report by Captain Bret Pliske, a Czech American intelligence officer, in October 1946. Alarmed, the JIOA requested that agents in Europe verify Pliske's report.[125] The FBI file on Eckert contains considerably more detail regarding the experiments, including an admission by Eckert of his participation.[126] The military's follow-up investigation, however, garnered sworn affidavits exonerating Eckert and claiming that no such tests occurred at the institute. Peter Pulz called the allegation "an infamous and malicious slander."[127] Eckert submitted a statement categorically denying the accusation. "As far as I know no such tests were conducted,"

Eckert wrote. "Furthermore there was no equipment available at this place which would have permitted such experiments."[128] Although this may have been true for the Prague institute, Eckert had also worked at a Luftwaffe research center in Brunswick where, according to the US military attaché in Prague, "[i]t is possible that experiments on human beings were made there."[129] Colonel Donald Putt considered the case closed once he determined that the source of the accusation, Pliske, was no longer assigned to AMC.[130] Pliske informed the FBI that "he had the feeling but could not prove that the German scientists at Wright Field were withholding some information in connection with their work."[131] Putt, an indefatigable advocate for his Germans at Wright Field, distanced himself from Pliske's report. Without verification of Pliske's accusation, which no one in JIOA or AMC truly desired, Eckert's immigration proceeded unencumbered.

The Question of Japanese Scientists

A "great Japanese inventive mind" contributed to the war effort even in the face of acute shortages and incessant bombing. Free from having to coordinate with other Allies and share occupation duties in Japan, the United States chose to survey and exploit Japan's war industries in theater rather than initiate a Japanese version of Paperclip, in which significant numbers of scientists and technicians would travel to the United States for extended periods. The Supreme Command for the Allied Powers (SCAP) enforced a similar regime in Japan to the one OMGUS instituted in Germany, specifically the strict prohibition of any research and development with military applications. Japan's engineers, perhaps as many as ten thousand, were mostly jobless by spring 1946.[132] Imperial Japan marshaled its talent and resources into conventional weapons production because it faced an insurmountable production gap compared to the United States. "Science in Japan was organized," one September 1945 memo began, "but it was organized behind existing industries to solve the immediate development production problems of these industries, particularly in regard to providing those industries with substitute materials to relieve critical shortages which were many."[133] Still, Japanese aeronautics proved exceptional in several categories, and US investigators took note, securing both prototypes and associated personnel with the same speed and thoroughness as the T-Forces had in Europe.[134] On at least one occasion, SCAP informed AMC that a particular Japanese scientist might be of interest to ongoing research projects. A March 1947 memo recommended that Takashi Shima, an aeronautical engineer who worked on a long-range bomber, be sent to Wright Field "under the same project as . . . the German Scientists now practicing in our laboratories," arguing the "investigation will bear fruit if followed up from several departments."[135] After careful vetting by the AAF, however, Shima was "not considered to be properly qualified for exploitation by Army Air Forces."[136]

Fears raised by U-234's unusual cargo spurred the creation of an Atomic Bomb Mission to scour Japan for evidence of progress toward a nuclear weapon. The assigned units impounded every ounce of uranium in use in labs, universities, and government-sponsored projects as well as monitoring any publications detailing experiments. Investigators swarmed the major universities, interrogated anyone working in nuclear physics, and established a surveillance regime that endured into the 1950s. SCAP soon discovered it had little to fear and even refused offers for extra personnel to staff the unit. By May 1946, SCAP analysts determined that "[t]he level of research in nuclear energy development in Japan had not advanced beyond the university experimental level," and that the Japanese did not even have a pilot program for separating isotopes. SCAP shifted its resources away from intensive surveillance and returned an analyst from the Manhattan Project, noting that "the task of exercising surveillance over existing industrial facilities does not require the expert knowledge."[137] The Atomic Bomb Mission first secured Japan's uranium stock and determined its progress toward a bomb was "negligible" before enlisting Japanese help in collecting readings and medical data from the bomb sites of Hiroshima and Nagasaki.[138] SCAP eventually encouraged Japanese scientists to stay abreast of developments in nuclear physics and virtually every other field, specifically by assisting exchanges and short-term residency abroad.[139]

One area in which the United States had considerable interest was Japan's biological warfare program. The infamous Unit 731, the Imperial Japanese Army's biological warfare unit, conducted horrific experiments on living subjects in Manchuria. The US government provided immunity from prosecution to Lieutenant General Ishii Shiro and other key figures from Japan's biological and bacteriological warfare initiatives in exchange for access to their research.[140] SCAP was tasked in May 1949 to compile a report to the JCS "on the material currently available . . . concerning biological warfare potential in Japan. Such a report will consist of details concerning outstanding biological labs, research workers doing work in sensitive fields, and typical projects being worked on."[141] The request mirrored the Atomic Bomb Mission, but unlike the survey of nuclear material and specialists, the United States clearly seems to have hoped to use the results of Japan's extensive work and experimentation for its own weapons program.

As Unit 731 demonstrates, the US military did not dismiss Japanese science, but SCAP decided the limited number of Japanese targets compared with Germany did not warrant a parallel Paperclip. Citing language barriers and poor intelligence, historian Takashi Nishiyama writes that the United States knew less about Japanese science than German going into the occupation. Furthermore, Japanese engineers faced economic and social pressure to remain home and care for extended family amid the ruins of the empire, whereas German scientists frequently worked abroad and could rely on a largely assimilated German immigrant community in the United States. The Japanese, in stark contrast, obviously faced deeply held suspicion and racial animus from Americans.[142] The national security bureaucracy, aware of its relative ignorance of Japanese science, reserved the right to pursue a Japanese scientist program in the future, but most assessments cautioned against it.[143] In a September 1948 memo entitled "Utilization of Japanese Scientists by the United States," H. C. Kelly, chief of the Fundamental Research Branch in SCAP's Scientific and Technical Division, recommended that the United States avoid becoming embroiled in a Paperclip-type scenario. Kelly had clearly followed Paperclip's trajectory and recommended that any Japanese scientist traveling to the United States be sponsored by "some civilian agency who would be responsible for security measures, rather than to import them into the US as was done in Germany. The adverse publicity, even by American scientists themselves, due to importation of the German scientists has proved our method of approach to be more feasible."[144] Some sympathetic SCAP officials advocated sending Japanese scientists abroad to provide opportunities to work with advanced equipment and improve their skills. Echoing arguments generated by Paperclip advocates, I. Rabi from the US Scientific Mission to Japan believed the pool of underemployed Japanese scientists could alleviate "the shortage of highly trained personnel in the US."[145] As tensions mounted in Asia, however, the need to resuscitate Japan's economic and military potential overruled these considerations.

Allied Exploitation Efforts

The United Kingdom and Commonwealth

The close relationship between Britain and the United States influenced each nation's efforts to exploit German science and scientists. Anglo-American intelligence cooperation benefited both parties and certainly contributed to victory in Europe and a more efficient occupation. CIOS, which developed the black lists of German scientists and technicians, eventually divided into FIAT in the US zone of occupation and BIOS (British Intelligence Objectives Subcommittee) in the adjoining British zone, although the two services continued to communicate with one another and share information despite sometimes competing for the same scientists. British military officers, like their US counterparts, recognized that snatching advanced equipment from Germany's research institutes and factories constituted reparations. Lt. General Sir Ronald Weeks, the deputy chief of the Imperial General Staff, admitted that capturing equipment and personnel was "one of the most vitally important of our immediate postwar aims . . . it may be this is the only form of reparation which will be possible to exact from Germany." Project Backfire required a joint effort, but the United States absconded with the most prized scientists and the bulk of the remaining V-2s under Allied control. The British lacked resources, and many of the Germans contemplating contractual work preferred the United States to the other Allies. "We despised the French," explained one scientist explaining the postwar dilemma, "we were afraid of the Russians and we didn't think the British could afford us. That left the Americans."[146] German scientists worked for every power, usually voluntarily and for various motivations, but Paperclip's success depended in part on the Germans' perception of American wealth and indifference to Nazism.

Despite scarce resources, the British aggressively pursued an exploitation program focused on German military research and industrial processes. The Defence Scheme, created in concert with US officials assigned to the Joint Staff Mission, apportioned Germans to military projects in Britain.[147] The British Chiefs of Staff informed the United States that "[t]he British view is that German science and technology should be exploited as fully as political and security considerations permit, in the interests of defence research in the UK and the US in order to develop our military potential at Germany's expense."[148] The British equivalent of Overcast, Operation Matchbox, involved consulting a limited number of German scientists and military experts. The Royal Air Force (RAF) targeted Germany's aviation industry in Operation Surgeon, which arranged for the transfer of the Focke-Wulf experimental lab to Britain, and Operation Medico, which secured equipment for a new RAF college.[149] The RAF was especially impressed by German wind tunnels and sent several missions to Völkenrode to acquire as much knowledge as possible. The British conducted 726 visits to the American zone between January 1946 and January 1947, indicating a high degree of cooperation.[150] By early March 1946, some two hundred German scientists and technicians worked in Britain, mostly for military projects.[151]

The Darwin Panel Scheme offered German scientists to British industry, provided they cleared US authorities.[152] British officials were sensitive to public criticism "if remunerative employment were given to ex-enemy aliens" and assured citizens that "in no case would German experts be employed in positions which might otherwise have been filled by British subjects."[153] In truth, British "employment" of Germans in private industry was often rather draconian. Operation Bottleneck extracted trade secrets from Germans, some of whom were neither scientists nor affiliated with the Third Reich. A British official recalled the procedure: "An English manufacturer would name his German counterpart and competitor and 'invite' him to England (whether the man comes voluntarily or not is questionable). They then discuss business and the German is gently persuaded to reveal secrets of his trade. When he refuses, he is kept in polite internment until he gets so tired of not being able to return to his family that he tells the Englishman what he wants to know. Thus for about 6 pounds a day the English businessman gains the deepest secrets of Germany's economic life."[154]

The United States condemned the Soviet Union's unilateral removal of German equipment and the "kidnapping" of scientists and their families from its occupation zone, mostly because of the loss it represented to Paperclip, but British occupation authorities behaved in a similar manner. A British military memo from August 1946 described how to approach a targeted scientist: "Usually an NCO arrives without notice at the house of the Germans and warns that he will be required. He does not give him any details of the reasons, nor does he present his credentials. Sometime later the German is seized (often in the middle of the night) and removed under guard."[155] The CIC reported on several cases in the Soviet sector of occupied Germany that closely resembled this scenario.

The British intensified its German scientist programs in the wake of Soviet aggressiveness in its own occupation zone. Like the United States, the British invoked "denial" when justifying contracting unsavory Germans with Nazi Party credentials. When British scientists and others in the public raised objections, the military made use of the Commonwealth by storing Germans in Australia, Canada, and even India. Some fifty-five German scientists moved to Canada before October 1950 and were given temporary migrant status. Canadian immigration authorities relaxed the prohibition against immigrants who served in the German army or had a "nominal" connection with the Nazi Party. In March 1987, Canada's Commission of Inquiry on War Criminals condemned Canada's participation in Operation Matchbox because several of the immigrants "had been involved in the Nazi war effort."[156] Australia accepted at least 127 German scientists and engineers between 1946 and 1951. A reported 31 were Nazi Party members, including a senior Nazi assigned to occupied Poland and IG Farben experts. Some Germans remained in Australia working on defense projects like guided missiles, atomic research, and advanced aeronautics.[157] Both the United States and Britain relied on friendly powers to house some of the "hot Nazis" in their respective exploitation programs, a secret that took decades to surface.

The ACA in Germany administered the mutual exchange of intelligence relating to Germany, including sharing information on scientific and technical data, but neither British nor American representatives honored this agreement with the Soviet Union.[158] Anglo-American intelligence offices exchanged information about targets on the black list and shared lessons learned concerning the treatment of scientists and dependents. An American officer recommended the British refrain from treating scientists as prisoners, for example, noting that when the Americans tried this at Wright Field, "the results have not been particularly happy. There is a definite air of restraint about the whole proceeding; segregated housing and working with military guards always about is certainly not conducive to productive working and thinking." Moreover, the officer related the problem of separating scientists from their "normal life, their records, and their assistants who may well have some of the essentials on their fingertips."[159] Anglo-American intelligence services coordinated efforts to secure each other's scientists in their respective zones and to provide "extra amenities" to contracted scientists and their families, but giving scientists preferential treatment in the occupied zones alienated the rest of the population.[160] OMGUS warned, "The moral effect on the other citizens and the lessening of respect for Military Government and the local German government will do much to lessen the value of the instruction in true democracy which we are endeavoring to foster." OMGUS recommended that British and American authorities arrange for the wholesale transfer of scientists and dependents from Germany to staff "special projects."[161] By early 1946, after months of trial-and-error handling, German scientists in captivity, clarifying their status, evaluating their expertise, overcoming legal and bureaucratic obstacles, and assessing Soviet intentions in Europe, the Western Allies moved toward a policy of long-term exploitation in their home countries.

France

France's disregard for the integrity of the zonal boundaries and its aggressive "recruiting" of scientists for its own military research exasperated American officials as much, if not more so, than the Soviet Union's myriad violations. In spring 1946, Lucius Clay and Secretary of War Patterson agreed that the French, not the Soviets, were the greatest impediment to a united occupation.[162] By 1949, according to one intelligence estimate, the French employed the second largest number of German scientists and technicians (one thousand) after the Soviet Union (ten thousand). The United States transferred approximately five hundred scientists across the Atlantic under Paperclip.[163] US civilian and military officials were vexed by French intransigence and brazen actions, ranging from, according to a June 20, 1946, telegram, "removing machinery from French zone in Germany without regard to reparations agreements" to outright kidnapping.[164] A week earlier the military had informed Robert Murphy, the US political adviser for Germany, "We have caught the French red-handed again stealing scientists out of our zone, and I am going to send you shortly some facts and figures on the subject in case you wish to discuss the matter."[165] With the CIC devoting the bulk of its resources to targeting the Soviet zone, largely with the aid of German POWs and ex-Nazis in its service, the French seemingly ransacked the Western zones at will.

US intelligence officers tracked French efforts to secure key scientists and weapons systems as early as December 1944. Herman Mayr, a German radio engineer, told the OSS that the secretary of the French Committee on Industrial Production ordered Mayr "not to reveal to the Americans or British any details of German radio and radar techniques, manufacturing methods, and research." The director of Askania, a German firm working on a homing torpedo, "stated that he had been instructed to reveal his work only to the French Navy. The French told him that to reveal details of his work to the British or Americans would result in his automatic inclusion on the Allied list of war criminals."[166] Still, the French thought little of harboring known war criminals such as Otto Ambros, the chemical warfare expert sought by Anglo-American interrogators. The CIC noted that French authorities, like the Soviets, recruited by exhibiting a "[c]omplete disregard for political pasts of scientists."[167] Ambros accepted French aid in escaping the US zone to work as the manager of the IG Farben plant in Ludwigshafen. Karl Rawer, another German employed by the French, supposedly directed "a ring which was providing scientific personnel and material for the French" originating in Freiburg.[168] Memos exchanged between the numerous intelligence agencies working inside occupied Germany reflect considerable angst over how to confront a supposed ally. The Enemy Personnel Exploitation Section (EPES) of FIAT asked FIAT leadership incredulously, "Is exchange of information on matters of interest to FIAT to take place between FIAT and the French authorities?" The July 19, 1946, memo cited yet another case in which the French arrested a scientist simply to avoid sharing the information with the Allies.[169] Days later, FIAT conceded that it "had no alternative but to hold up all requests to the French for access to German personnel."[170] In late May 1946, EPES complained "that the French have contacted and contracted several . . . Germans clandestinely," and despite signed agreements, "we [EPES] will continue to be very wary in giving the French any clearances to interrogate Germans."[171] British and American officials clashed over first rights to German scientists, but years of established protocols for intelligence sharing prevented truly damaging conflicts. The French were not part of this "special relationship," however, and it is not clear whether the activities inside the French zone were authorized or in any way organized by the French government.

The French usually did not resort to surreptitious recruitment because it was simpler for them to attract desperate Germans in need of immediate employment close to home. For example, the French dismantled the Dornier Aircraft Works in Friedrichshafen, which was in the French occupation zone, and rebuilt it in nearby Nay, France, under the auspices of the Turbomeca jet engine company. Employing over three hundred Germans, the French apparently began recruiting scientists for the enterprise in autumn 1945. "The area along the north shore of Lake Constance in the French Zone has been used as a collection and evaluation center for German scientists," American intelligence reported. "[T]he German scientists were gathered into teams; and, when the French felt that they had sufficient personnel in a given field, they were moved, with their families, to permanent installations in France."[172] Mirroring Soviet tactics, the French used respected and willing German scientists as recruiting agents. The French also made sure only positive reports reached targeted Germans, specifically pleasant living quarters, family reunion, and high salaries supplemented with special food rations. For all the consternation and agita the French prompted in US officials in Germany, the research conducted by Germans in French facilities was rudimentary compared to what Paperclip teams conducted in the United States. In some cases, the French even entertained inviting the United States to equip French labs with state-of-the-art machinery in exchange for research results, but the French apparently never formalized the request because of its "impracticability" and corresponding "political implications."[173] French subterfuge often frustrated US intelligence officers, but rumors of Soviet initiatives and plans for their captured personnel and equipment increasingly relegated the irksome French to an afterthought.

Securing advanced German technology and the brains behind it was an obvious priority for the Allies in spring 1945, especially with the war in the Pacific raging on with no end in sight. Anglo-American intelligence teams, AAF officers, US naval teams, the Red Army, and private commercial investigators descended on the Reich's disintegrating research and development infrastructure. The future of German science and the debate over what constituted reparations counted among the more intractable issues confronting postwar Allied planning, but every nation scrambled to secure what it considered the most valuable technology and personnel regardless of signed agreements. The military necessity argument dissipated with Japan's surrender, but a diverse cross-section of military officers, intelligence officials, senior civilian leaders, and representatives from the dozens of industries recast military necessity as "national security." Even without the Soviet Union replacing the Third Reich as a specter endangering US interests, the advent of the long-range rocket and the atomic bomb signaled a new age of warfare, requiring a vigilant and well-resourced MIC. Under these favorable conditions the JIOA targeted, located, and contracted the services of the best German minds available to supplement America's already robust defense establishment. The Paperclippers eventually assumed extremely influential positions inside the national security state, and the JIOA transitioned from recruiting scientists to expediting their immigration. The Germans became a permanent component of the MIC they served and, in no small part, created.

CHAPTER THREE

Conscientious Objectors

The State Department and Opposition to Project Paperclip

> I explained at several times, and at some length, that the State Department had a statutory duty to prevent the entry of people who would affect our security, and that it was not sufficient for us to get a certificate that a man was not a Nazi, or that he was a good scientist. What we are concerned with[,] I said[,] was the difficulty of assurances that the men in question would not take more than they gave and would not use the US as the place for continuing a technology which they would bring back to Germany at a later date.
>
> —Samuel Klaus, State Department representative to the JIOA

> Get that little Jew off the committee. . . . Get him off. He's a menace.
>
> —Major Simpson, JIOA staff officer

The exchange between Major Simpson of the JIOA and senior State Department official Herbert Cummings regarding the appointment of Samuel Klaus to the JIOA reveals the extraordinary personal animosity between the military's strongest Paperclip advocates and wary civilian bureaucrats opposed to an unfettered exploitation program. Samuel Klaus and others in the State Department, who had devoted their wartime service to preventing the Axis flight of capital and personnel abroad, understandably balked at a program designed to reconstitute German military research projects inside the United States. Klaus and his colleagues never opposed exploiting German scientists for their expertise, but rewarding them with US citizenship at the expense of thousands of displaced persons (DPs) languishing in central Europe was, in their estimation, both a moral outrage and an unnecessary security risk. The State Department's senior leadership ultimately conceded to the national security argument and paved the way for Paperclip's immigration strategy, but not before certain offices in the department confounded the JIOA and sponsoring agencies by insisting on enforcing existing statutes. The State Department did not exhibit the same intellectual dexterity and amoral pragmatism as some military officers when it came to reversing attitudes toward the German foe and the Russian ally. Although some opponents of Paperclip, Klaus most notably, were Jews offended by the prospect of bestowing Nazi scientists with contracts and citizenship, their opposition to Paperclip was rooted in an adherence to a worldview that still envisioned a viable Nazi threat, even after Germany's defeat. The State Department looked to the interwar period and questioned the judgment of those recommending immigration. What prevents a German resurgence other than a meaningful denazification and disarmament regime? Paperclip, opponents argued, undermined both objectives.

State Department resistance to the mass immigration of specialists without sufficient safeguards was grounded in the department's own conception of national security, which privileged international law and preventing the resurrection of Axis military designs abroad over a hyperfocus on the Soviet Union. While the military obsessed over a new enemy, seeking every advantage at hand, the State Department embraced its responsibility to prevent the resurgence of the old, especially in the Western hemisphere. The State Department was not the only bureaucracy to oppose Paperclip, but it posed the greatest obstacle to the program's success because of its legal responsibility for immigration matters. This chapter first details the initial conversations between the JIOA, senior civilian leadership, and the State Department regarding the wisdom behind the immigration strategy. If denial was the goal, civilian agencies argued, why not bring scientists over under "military custody"? The combination of national security and commercial interest in the scientists made immigration an attractive alternative to on-site exploitation in Germany or short-term contracts stateside. The next section explores the toxic relationship between members of the JIOA and Samuel Klaus. Klaus personified the State Department's wartime stance toward the Nazi threat as he worked diligently to prevent the resurrection of Axis centers of influence in the Western hemisphere. To Klaus, and to many of his colleagues assigned to Paperclip, abandoning this policy just months after the German surrender seemed ludicrous. Consequently, JIOA meetings were often combative affairs pitting most of the military representatives against Klaus, the designated conduit to the civilian bureaucracy, empowered to expedite the immigration program. Finally, the chapter concludes with the public opposition to Paperclip.

The military's public relations strategy, such as it was, failed to convince the public of the wisdom of transplanting one thousand Nazi scientists on to American soil. Critics ranging from Albert Einstein and Rabbi Stephen Wise to the Federation of American Scientists (FAS) and the National Association for the Advancement of Colored People (NAACP) excoriated Paperclip in the press and complained to the White House. While the public eventually grew indifferent to Paperclip, succumbing to the same national security logic as the rest of the bureaucracy, the opposition emanating from inside the government and from the outraged public represents an alternative perspective to the reigning logic of national security. For Paperclip's opponents, Nazism remained as much a threat as Soviet communism. Exploiting German science made sense, but inviting hundreds, if not thousands, of "ardent Nazis" into the country with minimal restrictions defied all reason, not to mention the law.

From Exploitation to Immigration

The agencies involved in handling the immigration component of Paperclip included the military (represented by the JIOA), the State Department and its consular offices, and the Justice Department, which controlled the Immigration and Naturalization Service (INS) and the FBI. The Commerce Department and private industry eventually joined the process and sponsored the immigration of dozens of Paperclip recruits. To hear the executive assistant to the attorney general H. Graham Morison describe the process nearly thirty years after the fact, Paperclip was a spontaneous and impulsive operation driven by the Soviet threat:

> [W]hen the question came up . . . of how in Hell we could grab the remaining German scientists that the Russians were systematically recruiting. How were we going to get them out of Germany. They wanted to come. You've got immigration, you've got all the bureaucratic things to do to do it promptly, and I said, "let me handle it," I said, "I'll handle it." So I called four members of the Little Cabinet. It was called "Operation Paperclip." I got the Air Force to get the planes. The immigration service is part of the Department of Justice, so I called the Commissioner of Immigration, over whom I exercised an oversight[,] to come over and told him what to do. We were going to bring these people into a certain place, where we had housing for them, we didn't want any "hurrah" after we got them here. The main thing was to get them here promptly and before the Russians got them. This was done within 24 hours, that's how we got von Braun.[1]

Morrison's flippant, somewhat self-congratulatory anecdote is not entirely inaccurate given the surreptitious exfiltration of the rocket team and dissembled V-2 parts to the United States. Convincing officials to overlook normal procedures for an extraordinary purpose, especially during the chaotic period after the Third Reich collapsed, did not obviate the need for a formal, legal, and consensual process. The State Department in particular pressed for an interim procedure "based on agreements between the responsible agencies of the government" dealing with questions of entry, responsibility for custody, the length of stay, compensation, restraints on activities, and assigning responsibility for implementing all of the above requirements.[2] Securing scientists and warehousing them inside the United States accomplished the immediate objective of denial, but determining the scientists' status was infinitely more complicated.

Few civilian officials questioned the logic of exploitation, even inside the United States, for a reasonable length of time, but this early agreement did not equate to support for immigration. Navy secretary James Forrestal wrote secretary of state James Byrnes in January 1945 about his intention to ship Germans over to "fill in some of the gaps in our research activities" and share the wealth with private industry.[3] Forrestal's letter prompted a correspondence between the War and State Departments during the final phase of the war in Europe and immediate aftermath, in which senior military officers gauged the State Department's willingness to assist the military with unfettered exploitation. In late June 1945, William Clayton, an assistant secretary of state, wrote Jack McCloy, his counterpart at the War Department, to signal the State Department's acquiescence: "We consider this transfer of specially qualified German scientists with appropriate provision for surveillance of their activities, to be fully justified on grounds of national interest." Clayton seemed particularly swayed by the denial argument, noting that if scientists are "domiciled here, we will be able to keep their knowledge and capacities from being utilized in many ways inimical to our interests."[4] Whether the State Department agreed with an exploitation program or not was immaterial; most senior officials read the writing on the wall. Green Hackworth, the legal adviser at the State Department, surrendered to the logic of expediency in July 1945, writing, "I assume . . . our government will not desist from using these processes whatever the legal consequences may be. In any event it is for the people who complain to establish the justification of their complaint on a supposed legal basis."[5] Many in the State Department had sufficient reasons to oppose immigration, legal and otherwise, but they were considerably less influential than those who either approved of Paperclip or regarded it as the military's prerogative.

The prospect of immigrating Paperclippers, up to one thousand of them and their dependents, exposed America's controversial quota restrictions on granting visas to Europeans. Ten million Europeans lived in horrid conditions as stateless DPs, principally the surviving remnant of European Jewry. Despite pressure from various corners of the bureaucracy and especially the public, only forty thousand people were admitted to the United States from Europe in the first three years after the war.[6] Allowing enemy aliens like the Paperclippers to fill the few slots available while the victims of Nazism languished in camps struck many as intolerable, including some responsible for implementing Paperclip. The State Department understood it faced blowback regardless of how it proceeded regarding Paperclip. If the department obstructed Paperclip, it risked endangering national security; if it caved to military necessity, thousands of deserving DPs continued to suffer while ex-Nazis took their place.

The State Department Visa Division mandated that scientists' dossiers include a personal history statement, including for dependents, as well as military records, political party affiliation, and contracts from the sponsoring agency.[7] Furthermore, the Immigration Act of 1924 required that applicants provide a "certificate of good character" approved by local police. Most Paperclippers and their sponsors found the requirements prohibitive because of the scientists' Nazi affiliations and, for the number of recruits who resided in the Soviet occupation zone, incomplete or destroyed records.[8] Those scientists transferred to the United States during summer 1945 qualified as "a privileged group of prisoners of war," as Clarence Lasby notes, but Truman's March 1946 authorization expanded Paperclip and legalized scientists' entry using two methods. The first and preferred method for the military involved securing an immigration visa for recruits, issued on a quota basis, and eventually having them apply for citizenship. The second method, granting a visitor's visa, allowed scientists to reside in the United States for six months on a nonquota basis, but citizenship was still the ultimate objective. One State Department employee commenting on the immigration process described the debates between agencies as a "long and sad story," noting that the "extralegal manner" placed the department "in a very difficult and embarrassing position" in terms of enforcing international agreements regarding anti-Axis efforts and admitting properly vetted immigrants.[9] The JIOA's role as the official sponsor of the scientists' immigration applications only exacerbated the State

Department's reluctance to sign off on the program, especially after the Visa Division discovered that the JIOA "revised" security dossiers at will.

Samuel Klaus was not alone in opposing the immigration of Paperclip specialists and the JIOA's reckless disregard for both denazification procedures and legitimate internal security concerns with importing hundreds of enemy aliens to American soil, but he certainly inspired the most vicious response. Klaus, a trained investigative lawyer who helped erect the BDC after the German surrender, seemed like a logical choice to represent the State Department on the JIOA.[10] Klaus is depicted as a lone moral crusader against an unscrupulous and deceptive military "conspiracy" in Linda Hunt's and Tom Bower's accounts of Paperclip and an obstinate, incompetent, and meddlesome bureaucrat in Clarence Lasby's uncritical narrative vindicating Paperclip as a Cold War victory. A careful reading of his memos and reminiscences reveals a more nuanced picture. Klaus was evidently a difficult personality throughout his career, alienating superiors and coworkers alike, but he brought considerable energy and purpose to every assignment. The JIOA expected a compliant, sympathetic bureaucrat to occupy the State Department chair on its committee—instead, it got Samuel Klaus.

Originally assigned to the Treasury Department's General Counsel Office and then the FEA during the war, Klaus played an important role in Operation Safehaven along with another future State Department colleague critical of Paperclip, Seymour Rubin. The brainchild of treasury secretary Hans Morgenthau, Safehaven sought to neutralize German industrial power and prevent the flight of German assets abroad before the Reich's collapse.[11] Klaus considered the mission vital to winning the peace and his job nothing less than counteracting German plans "to achieve domination of the world."[12] Klaus believed Safehaven and his career path inextricably linked. Consequently, he pushed for greater resources under his direction, reporting back to Treasury that "[t]he FEA must . . . have available to work under its direction in Germany hundreds of trained investigators" poached from the FBI, among others.[13] A December 1944 article entitled "Did You Happen to See Samuel Klaus?" described Klaus as "the movie-going, mystery story-loving, racket-busting, spy-catching super sleuth lawyer, who's now an assistant general counsel to the FEA."[14] Despite the positive profile, Klaus, who held the equivalent rank of brigadier general in the army during the war, was frustrated by the limited scope of the mission and subordinate role.[15] Klaus wrote his superiors in April 1945 "that there is an FEA-phobia in our staff here" and worried about "resignations, apathy, [and] distaste."[16] Klaus's discouragement likely influenced his subsequent interactions with the military, but he also revealed deeply held suspicions of Germans living abroad, including German American cultural groups, which he accused of "more devious . . . fifth columnism [*sic*]." Fully supportive of Morgenthau's draconian plans for a postwar Germany, Klaus warned against the dangerous combination of German wealth and expertise acting as Nazi "nest eggs" in neutral countries.[17] Klaus regarded the possibility of a "Fourth Reich" reconstituting on foreign soil a viable threat and threw himself into the Safehaven operation with characteristic zeal.[18] Asking someone like Klaus to suddenly reverse course and facilitate the legal immigration of people he once hunted was a predictable recipe for disaster.

When consulted as part of Klaus's security clearance for the Safehaven initiative, a colleague at the Treasury Department answered the question "What desirable traits are outstanding in the applicant?" with three words: "tenacity, trustworthiness, aggressiveness."[19] Klaus's enemies likely agreed on at least two of the three. Klaus's ambition and moral inflexibility sometimes provoked the ire of his superiors, even those who valued his commitment and expertise. This trend continued during his tenure at the State Department and contributed to his removal from the JIOA and eventual scapegoating during the McCarthy era. When Klaus traveled to Europe before starting his FEA assignment in spring 1944, Morgenthau exploded: "I made it perfectly clear . . . didn't I? . . . [t]hat I didn't know what he was going for, and that he would not go as a Treasury Rep, period. . . . What is he going over for, anyway?"[20] Morgenthau remembered "Sammy Klaus" as the guy who "is always trying to make trouble."[21] Klaus soon developed a reputation detrimental to his future assignment on the JIOA. He "made trouble" for military governor Lucius Clay, FIAT commander Ralph Osborne, and the Pentagon's Civil Military Affairs division chief Major General John Hilldring. The latter was especially portentous since Hilldring later assumed a senior post at the State Department and generally supported Paperclip. In a June 1945 memo discussing the fate of Klaus's FEA group, Osborne rejected the idea of placing him under military authority "[i]n view of the personalities involved."[22] Despite his irksome temperament, Klaus managed to secure valuable patrons like Morgenthau at Treasury and, later, undersecretary of state Dean Acheson. Klaus drove much of the State Department's opposition to Paperclip immigration, but his relatively junior position combined with a grating manner and, it must be said, the undisguised antisemitism within the government and military, ensured his dismissal from any function related to Paperclip.

Klaus may have held some sway with Acheson at first, but Acheson was ultimately the one who signed the memos and directives authorizing the expedited immigration program so vital to Paperclip's success.

Samuel Klaus best articulated the department's position on the exploitation program by responding to inquiries from inside the department and from the military concerning the requirements for immigration. Paperclip was unprecedented and required the active support of civilian and military agencies, but walking sponsoring agencies through a process that was essentially improvised challenged agencies burdened by the administrative responsibility of upholding the law. Klaus's memos reveal moral objections to Paperclip, but his primary concern was protecting the department from negative fallout. Klaus and some of his colleagues from Safehaven clearly doubted the Germans' intentions and talent, not unlike Walter Jessel did, but Klaus never undermined Paperclip from within or questioned the logic of denying scientists to the Soviet Union. Howard Travers, chief of the Visa Division in the State Department, agreed with Klaus and urged the military to retain custody of the scientists as long as necessary, arguing, "The first line of defense from the standpoint of security probably rests in the use of other means of admittance to the US than the issuance of quota immigration visas." Failing that, Travers notes, "the second line of defense dictates a very careful examination of the prospective immigrant prior to admission to the US."[23] Klaus and Travers expected Paperclip to involve a small number of "specialists whom it is desired to exploit in the military interest," not a pipeline of hundreds.[24]

While the bureaucracy debated expanding Paperclip to one thousand recruits complete with offers of citizenship for specialists and their families in spring 1946, Travers, with Klaus's assistance, drafted a lengthy memo summarizing his objections, beginning with self-evident conclusions—Paperclip violated the Safehaven program, JCS 1067, and virtually every regulation agreed to by the Allied Control Council (ACC). The military may be indifferent to international obligations, but the State Department assumed all the risk and responsibilities. Travers's memo could be read as a denunciation of the scientists and their advocates inside the military, but his declared purpose in writing it was "to point to those factors which will be most embarrassing if at a later date the policy of the departments concerned in this program is subjected to press scrutiny."[25] Travers argued, "To insure [*sic*] the Department of State the protection which it deserves because of the use of visa facilities, it should have the power of veto after the collection of all the information concerning the particular scientist."[26] Travers and Klaus's proposal was inconceivable given the prevailing logic of national security. The JIOA unilaterally froze the State Department, specifically Klaus, from the vetting process with the aid of sympathetic officials throughout the government. For all its publicly aired grievances against the State Department, the JIOA implemented Paperclip with minimal interference.

Klaus's hostility toward German scientists derived from his experience with Safehaven and alarming reports of Nazis abroad planning for an eventual resurgence. Klaus feared Germany would attempt to rearm secretly just as it had after World War I. Specifically, Klaus cited the precedent of scientists working for third powers like Sweden and the Soviet Union to perfect weapons and share knowledge. In a September 1946 memo, Klaus claimed, "German technicians and industrialists were briefed before the Nazi collapse on their postwar responsibility: to make certain of the survival of German military-industrial power and technical skill outside Germany." Klaus maintained that "Pan-German leaders counted on jockeying for position among the victorious allies to provide them with the milieu for their own renascence."[27] Like Walter Jessel, Klaus believed Paperclip rewarded the most charismatic and dangerous Nazis by providing them with their own haven on American soil, prompting the question he hoped every intelligence officer asked during recruitment: "Is the individual scientist likely to be a potential source of scientific knowledge which might be used against this country at a later date?"[28] Klaus consistently reminded his superiors that the State Department would be held responsible for allowing dangerous individuals inside the country, writing in a 1946 memo that "intelligence information now at hand has put the department on notice that many of these German scientists have been and may now be associated in movements to conduct espionage in the US, or to restore German military potential."[29] Once admitted into the United States, he reminded superiors, scientists were as free to travel as any other resident.

Aside from the obvious security fears, Travers and Klaus worried the United States could be "saddled with a number of 'drones' who by no stretch of the imagination qualify as outstanding scientists."[30] State Department official Seymour Rubin insisted on determining whether a particular scientist was absolutely necessary for the program. Rubin complained that army officers in Germany "wanted to bring the guy who soaked the floor in the rocket establishment on the grounds that they would contribute to our own efforts."[31] Rubin, Travers, and Klaus urged bringing specialists over on probationary status. "Otherwise," Travers argued, "the United States government has traded all of its trinkets and cannot force compliance with any demand." Military custody denied the Soviets and preserved the law while enabling military projects to proceed immediately. Incentives came with time, Travers posited, concluding his spring 1946 memo with the rhetorical question, "Why give them everything before they deliver anything?"[32] Samuel Klaus informed his counterparts in the military that the State Department would not "confer immigration status as a 'sweetener'" for accepting a Paperclip contract. "This is a War Department responsibility," Klaus wrote, "and the War Department should provide the 'sweetener' out of its own authority."[33]

State Department officials dedicated to securing the Western hemisphere from Nazi influence were understandably appalled by a program inviting Nazi scientists to continue their forbidden work inside the United States. Spruille Braden, the assistant secretary of state for American republic affairs, had labored over agreements preventing nations like Brazil, Argentina, and Mexico from harboring Germans during the war. Like Klaus, who had worked in Braden's Latin American Affairs Office before the JIOA assignment, Braden feared the Nazi hydra could rear its ugly head again if American republics, for their own enterprises, recruited Nazis on the run.[34] It never occurred to Braden that the United States would be the first to breach the international agreements he had negotiated. In May 1946, Braden reminded Acheson of the "adverse repercussions" should the United States pursue Paperclip, since the program violated inter-American agreements "directed against the reconstruction of Axis 'centers of influence' in the postwar period."[35] Citing resolutions adopted by the 1945 Mexico City Conference, Braden reminded Acheson that dozens of governments agreed "to prevent Axis-inspired elements from securing or gaining vantage points from which to disturb or threaten the security or welfare of any Republic." Braden expected backlash from several Latin American countries for demanding they expel German scientists and technicians while the US government happily facilitated their immigration to the United States. Braden understood "the burden would fall on the department to explain this program to the American public" and joined State Department critics advocating scientists be brought over as POWs to be "milked of their knowledge" and returned.[36] Samuel Klaus also cautioned Acheson about the provisions embedded in various inter-American treaties, noting the potential embarrassment of revealing "our own large-scale violation" of provisions the United States had pledged to implement. Paperclip, Klaus continued, "covers persons whom unquestionably our international commitments require us to expel from this country and from every country in the hemisphere."[37]

Spruille Braden was especially concerned about Juan Perón's Argentina reaping the fruits of the defeated Third Reich. Braden wrote a provocative article in the April 1946 edition of the *Atlantic Monthly* condemning Perón for his overt courtship of ex-Nazis to coincide with his release of the eighty-six-page Blue Book denouncing Argentina for aiding Germany during the war and subverting neighboring republics.[38] Braden detailed Argentine parroting of Nazi culture and symbols and the extensive contact between the two regimes' military and intelligence services, concluding, "As long as there survives the possibility of a renewal of military aggression, the German politico-economic system in America will be full of dangerous meaning for those of us concerned with the security of our families and our neighbors' families."[39] Braden had reason to worry, as thousands of displaced Germans with ties to the Nazi regime voluntarily migrated to Argentina. Adolph Galland, the renowned German fighter ace and personal friend to Perón, recalled that "the idleness of many German scientists, technicians, and other specialists after 1945 suited Argentina very well." Furthermore, Galland believed the Germans appreciated "serving a nation that approached them sympathetically and without prejudice and serving their own fatherland at the same time."[40] The CIA validated Braden's prediction of a Nazi revival in a 1953 intelligence report confirming that a "substantial stream of German immigration, including Wehrmacht and SS veterans, Nazi economists, propagandists, intelligence agents, scientists, and military specialists, has flowed into Argentina since 1945."[41] That the United States would indirectly aid Perón's fascist regime by expanding a controversial program like Paperclip violated all logic and mocked Braden's personal investment in his wartime mission.

The JIOA sought to expand the potential pool of exploitable Germans by encouraging immigration to "Brazil and other allied nations of the Western Hemisphere" in a proposal both Braden and Klaus regarded as a thinly veiled attempt to circumvent existing immigration laws.[42] Klaus wrote Braden that the JIOA plan would either "be limited to anti-Nazis and others who are needed to maintain a denazified German economy and technical civilization, or to Germans who are unwilling voluntarily to go to the US or whose security record is so bad that they would not be eligible." Klaus foresaw "considerable competition between the US and elements in the other American republics for voluntary immigration with political effects that are bound to be undesirable."[43] Klaus and Braden formed an effective team, convincing both Acheson and General John Hilldring, who left the Pentagon to become assistant secretary of state for occupied areas, that the JIOA proposal risked damaging the department's reputation abroad and with the American public, since "[i]t is quite likely that the persons recruited would include many politically questionable persons, in excess of the 1,000 on the present list for military exploitation in the US, whose presence in the Western Hemisphere would violate the language and spirit of the inter-American commitments."[44] The JIOA claimed the army, the responsible investigative agency in Germany, "will not permit 'ardent Nazis' to leave Germany for other nations in the Hemisphere."[45] In practice, the JIOA cared little for discerning the ideological proclivities of Paperclip recruits, let alone German scientists working outside the United States. Hilldring decided that the State Department would continue to decide each immigration case individually based on a complete security evaluation.[46] Klaus and Braden's success was fleeting, however, as the JIOA succeeded in transferring several controversial Paperclippers to "friendly" governments indifferent to toxic ideology or war crimes. The Visa Division reported in March 1948 that several Latin American nations, including Mexico, "may have been promised by the military authorities a share in the number of scientists to enter that country for the purpose of obtaining immigration visas with which to reenter the US, unless Mexico is allotted a number of the scientists which they may desire."[47] This quid pro quo arrangement continued well into the 1950s.

The State Department's opposition to Paperclip rarely extended to the most senior posts. Klaus and his colleagues working in the Visa Division, the Office of Controls, and Latin American Affairs frustrated the JIOA for a time by demanding the agency adhere to existing laws, but Secretary of State Byrnes, Acting Secretary Acheson, and Secretary George C. Marshall supported an expanded Paperclip program and immediately simplified the vetting process for sponsoring agencies. While Klaus battled the JIOA over altering documents and ignoring established protocols, more senior State Department officials assured their counterparts at the War Department that an expedited process was forthcoming. Assistant secretary of war Howard Petersen relayed details from a meeting with Hilldring over immigration, noting that Byrnes and other senior executives "regretted that administrative procedure within the State Department had thus far blocked the effective implementation of the agreed . . . policy." Petersen was confident the State Department would "concur in a War Department proposal to bring over 1,000 German scientists and their families under military custody, with a provision that visa arrangements might be made later for those whom we would wish to keep."[48] The military stressed the urgency of removing scientists from Western Europe over establishing their legal status. Both the War and State Departments preferred postponing complex matters like immigration to focus on recruiting and retaining needed specialists.

Dean Acheson, the acting secretary of state in summer 1946, listened to Paperclip critics and even objected to some military proposals, but he ultimately agreed to expand Paperclip, knowing citizenship was the likely outcome. Acheson digested intelligence assessments from Europe and concurred with the military's assessment that denial to the Soviets warranted extraordinary measures. In an August 30, 1946, memo, Acheson urged President Truman to approve the JCS proposal to relax custody arrangements and allow family members to join the recruits, "[s]ince cooperation of the specialists is necessary to successful exploitation." Concerned about the potential blowback over such a controversial policy, Acheson reminded Truman that the "War Department would be responsible for custody and for excluding from the program persons with Nazi or militaristic records."[49] Acheson did not commit the State Department to approving citizenship for every candidate the military nominated, but Klaus and others whose job it was to enforce immigration requirements found it exceedingly difficult to refuse a JIOA-sponsored application once the scientist and his family were already on US soil.

In August 1947, a few months after Klaus was sacked from the JIOA, George C. Marshall confirmed the State Department's full cooperation on the immigration issue to secretary of war Kenneth Royall. Marshall, after fielding numerous complaints about Klaus and others lower in the bureaucracy, acknowledged that "the present war-time restrictions superimposed upon the admission of aliens under the general immigration laws and regulations constitute an impediment to the operation known as Paperclip" and agreed to "exempt the aliens in question from certain provisions . . . upon the basis of a certification by the Secretary of War, Secretary of Navy, or the JCS, that the admission of each such alien is 'highly desirable in the national interest.'"[50] Samuel Klaus's understanding of what constituted the "national interest" contradicted the military's definition. With the military dominating the national security agenda, the State Department effectively joined the cast of supporting players in Paperclip after abdicating responsibility for determining who among the hundreds of German scientists qualified for citizenship.

Samuel Klaus versus the JIOA

Those in the State Department opposed to an unrestricted Paperclip were motivated in part by a healthy suspicion of the German scientists, but most acted primarily to shield the department from assuming the risks inherent in Paperclip while receiving none of the benefits. Paperclip also exacerbated the growing rift between the State Department and an ascendant national security bureaucracy increasingly dominated by military officers entrenched in the JCS, the sponsoring military services, and even the State Department. Samuel Klaus was vilified by officers resentful of a man they perceived as a meddling civil servant ignorant of the new world the United States inhabited after 1945, but Klaus's personal and reflective memos contradict this depiction. Klaus did not dispute the logic of military necessity, even encouraging the JIOA to secure thousands of potentially useful specialists under a military custody program, but he fiercely protected the State Department's prerogative concerning visas. Klaus's superiors agreed, specifically Hamilton Robinson, the director of the Office of Controls, who shared responsibility for Paperclip within the department along with General Hilldring. According to Linda Hunt, Robinson, an associate of future secretary of state John Foster Dulles and the legal adviser to the SWNCC, "was . . . also a carbon copy of Klaus in his suspicions of the military officers."[51] Howard Travers, the chief of the Visa Division, and Herbert Cummings, the assistant chief of the Bureau of Foreign Activity Correlation, joined Klaus and Robinson's outrage concerning the military's dismissive attitude toward existing regulations governing immigration.[52] "It should be recognized that any concessions with regard to security standards," Cummings wrote Robinson, "will ultimately reflect not upon the War Department but on the State Department—both in the public mind and in the matter of ultimate legal responsibility."[53] The venom directed at Samuel Klaus personally is indicative of more than Klaus's grating personality. The JIOA viewed Klaus's obstruction as a stubborn remnant of an outdated wartime mentality. The national security state thrived on consensus over the Soviet threat, and while Klaus conceded the point, he believed Nazism remained a threat.

Klaus's embattled tenure with the JIOA began in spring 1946 with his appointment by the State Department's Intelligence Office. Fortunately for historians, Klaus made a habit of transforming copious notes taken during JIOA meetings into extremely thorough memos for the record. Years later, when the State Department and Klaus personally contended with the likes of J. Edgar

Hoover and Joseph McCarthy, Klaus's memos proved an effective counter to charges that the department undermined Paperclip purposefully. Klaus managed to deflate the JIOA's expectations for a pliable rubber stamp at the very first meeting he attended on April 24, 1946. Klaus delineated the State Department's skepticism of the immigration program and soon learned that "[t]here seemed to be considerable confusion among the people present as to just what the State Department could do to compel entry of the scientists selected by the War and Navy departments."[54] The JIOA assumed the matter was settled and that using the existing quota to accept scientists over DPs should proceed unencumbered. Within hours of the meeting's conclusion, Klaus learned that Cummings "had received a call from G-2 complaining and asking if the department could not send someone else." Klaus took the slight in stride, informing Cummings "in the event of a future call it would be well to say that the department would carry out the agreement and that there was no division of opinion on that subject but that we must be concerned with the problem of security."[55] Klaus, contrary to the JIOA's interpretation of events, did not sabotage Paperclip from the outset. As a State Department representative, Klaus prioritized protecting the department over enabling the JIOA's evasion of the law.

Two days after the April 24 meeting, Klaus drafted an internal memo addressing "State Department problems," in which he demonstrated a willingness to assist the JIOA without violating the law, treaties, or the internal security of the United States. The memo proves Klaus was not the unreasonable, subversive rogue employee bent on thwarting national security out of some misplaced moral crusade. This excerpt states succinctly both Klaus's and the State Department's position:

> Regardless of the decision, the department must observe its commitment "to facilitate" the bringing in of such scientists. However, that facilitation must be consistent with department statutory obligations respecting the issuance of visas, and with the treaty obligations, . . . and with its duties to the American public. It would seem that for the protection of the department from reasonable criticism now, or in the future, the department should insist upon a maximum of security protections in the selection of immigrants, the investigations of relevant factors, the nature and conditions of employment in the US, the surveillance of observation of the immigrants, and the imposition of such controls over their behavior, their movements and their activities into the indefinite future as to make a record that the State Department insisted upon the

maximum of protection against all those security and other evils which the State Department is required or expected to guard.[56]

The more Klaus interacted with the JIOA, the less convinced he became that the agency respected the department's concerns and obligations. Klaus distrusted German scientists instinctively because they had substantive ties to the Nazi regime, and he soon lost confidence in the JIOA's capacity and willingness to observe even minimal security standards.

Klaus operated under the assumption that while Paperclip was an extraordinary venture, assisting the program did not preclude sacrificing security for expediency. If the Soviets snatched scientists left and right, as the JIOA claimed, military custody accomplished the same objective as immigration, only faster. In the April 26 memo, Klaus reiterated his apprehensions about immigration as a solution to denial, noting that a permanent immigrant "must intend to leave everything German behind him and to make the US his home forever and ultimately to pledge undivided allegiance to the US, repudiating all German connections." Understandably suspicious of the Paperclippers' intentions, Klaus recommended they "be admitted only on a lesser basis, such as a temporary visa for limited purposes, under close surveillance."[57] The memo reminded his superiors that once citizens and therefore free to move and travel at will, scientists could transmit their skills and knowledge acquired in the United States to Germany or the Soviet Union at "a propitious moment." Furthermore, Klaus predicted accurately that "the influx of scientists in the US creates a burden on the FBI and other internal security agencies."[58] At no point did Klaus condemn Paperclip outright or suggest the department should stymie the effort, but the JIOA saw it differently. The JIOA leadership expected compliance, not input, and subsequent meetings only aggravated the fissure between the military agency and the State Department.

Personal animosity between Klaus and JIOA committee chairman Bosquet Wev, a US Navy captain deeply invested in Paperclip's success, dominated practically every JIOA meeting between spring 1946 and 1947. The first issue involved agreeing on the chain of command and sharing information. Wev regarded the JIOA as a clearing house for intelligence, targeting, and document collection. Paperclip was vital to America's evolving national security concerns, and the more scientists the JIOA could identify, secure, and exfiltrate from Europe before the Soviet Union did the same, the better. Wev's intransigence and dismissive attitude toward rules and regulations reflected the War Department's privileged position, a point he reiterated frequently in meetings. That Klaus would question the façade after just one week by creating "legal obstructions" infuriated the military contingent that made up most of the JIOA staff.[59] In the April 26, 1946, meeting, Klaus "argued that it seemed to me that a coordinating committee of the high level of this one was expected to do more than just a clerical job, but this argument did not impress the other members."[60] Klaus further frustrated Wev by rejecting his ploy to grant the JIOA chairman the "authority to act without concurrence of the Governing committee in certain cases."[61] Klaus responded by noting that Wev's proposal effectively sidelined the State Department among others from decisions affecting its mission. "I pointed out that so long as JIOA existed under its present directives," Klaus recalled, "any act disregarding State Department representation was especially illegal." Klaus's objections worked temporarily. "It was finally agreed to limit his [Wev's] authority to executing administrative matters," Klaus reported, but Wev remained obligated to inform the committee of any contacts related to the exploitation program.[62] Klaus's small victory infuriated Wev and escalated the dispute.

Klaus quickly relayed his concerns about the JIOA to his superiors in the State Department. "I have become convinced that there is a grave danger of embarrassment to the department and to the government, nationally and internationally, in the present uncoordinated handling of this problem," Klaus wrote Joseph Anthony Panuch in the Administration Office.[63] Klaus was especially alarmed at Wev's nonchalant attitude toward the security dossiers, noting that "Wev indicated that he did not believe that the Services would be in a position to comply" with the required visa paperwork. "Superficially, he [Wev] expressed lack of confidence that any substantial investigation would be made in the US for lack of personnel."[64] The CIC in Europe struggled with a backlog of requests for investigations, and the Soviet threat consumed most of its resources, but Klaus spoke for General Hilldring and Howard Travers when informing the JIOA "that the Army would be compelled to introduce a better CIC system if we were to rely on army investigations for visa issuance."[65] The JIOA's haphazard intelligence network in Europe endangered Paperclip's immigration proposal, at least from the State Department's perspective. Spruille Braden, echoing Klaus, informed Dean Acheson, "[I]t appears that the Army is not supplying the department with full information . . . to the effect that large numbers of so-called German scientists who have come to the US under the military program have turned out to be so mediocre technically that they are actually being sent back to Germany."[66] Herbert Cummings appreciated Klaus's diligence and corroborated his suspicions regarding the broken investigative process: "State Department officials in Germany on many occasions during the past two years have been made aware of the War Department's failure to examine even the most accessible pertinent records such as the Nazi party files at the G-2 Document Center in Berlin."[67] The JIOA pressured the civilian bureaucracy to accept its investigations of Paperclip recruits without question, knowing full well it was not equal to the task. Hamilton Robinson aided the beleaguered Klaus by confirming that his office would not act on an application until all outstanding questions were answered. Robinson promised the JIOA a "statement of deficiency" and the opportunity to resubmit.[68] But the JIOA simply "revised" dossiers on its own accord.

Klaus insisted on precision of language when defining terms and a scientist's status, but the JIOA used ambiguous language as a cover for some of their more problematic recruits. In comments regarding the so-called Gossett Bill, designed to prevent the immigration of dangerous enemy aliens, a policy Klaus obviously supported, he challenged the inexact terminology contained in the bill. "What is 'Nazism' and what is 'fascism'? Is it limited to the National Socialist Party, or to the Italian fascist party, or is it intended to cover a concept? If a man was in the SS or SA or Gestapo, can he become eligible for entry by showing that he preferred, at any time between the dates, some other totalitarian leader than Hitler or Mussolini?"[69] Klaus utilized this incisive questioning during his JIOA assignment as well, much to the displeasure of the military staffers seeking the swiftest path to citizenship possible. Klaus worried that educated and refined scientists concealed their ideology and criminal complicity from interrogators. "It is unlikely that scientists would be street brawlers or concentration camp managers, or otherwise active in the cruder aspects of Nazi political activity," Klaus wrote Acheson. Klaus could not fathom that a privileged scientist coddled by the Nazi state would somehow abandon their beliefs just because they worked on "American soil."[70] Klaus failed to convince the rest of the JIOA that issuing visas "without any question and regardless of evidence" endangered Paperclip's viability. "That the Department's security fears were not baseless," Klaus wrote shortly after the Georg Rickhey affair, "was recently demonstrated when a war criminal, wanted for crimes of a bestial kind, was found here among these scientists."[71] Klaus had more than enough evidence to support his well-founded criticism of immigration, but few in positions of influence believed such revelations warranted suspending the policy.

The most infuriating aspect of Klaus's tenure with the JIOA was his inability to convince officers like Wev that he shared their primary objective.

Klaus deeply resented the accusation that he opposed exploitation, and therefore undermined national security. A JIOA officer told Klaus that he "understood the State Department's position to be that we laid a greater emphasis on the danger of a German comeback than on the Russian threat." Klaus responded that the officer was wrong, "that we had in mind both problems—the German and the Russian—and that we desired a solution that would take care of them both."[72] If the JIOA blamed the State Department for leaving scientists vulnerable to Soviet exploitation, voluntary or otherwise, by insisting on adhering to existing requirements, why not simply detain them? "If it is not desired to keep such Germans in strict military custody," Klaus wrote Acheson, "they should be sent to some other place where their activities and movements could be completely controlled, or they should be left in Germany and more strictly controlled there, an obligation which belongs to the army."[73] Klaus believed President Truman's order to admit "a limited few" scientists whose work "would promote scientific research in irreplaceable ways and would be a bona fide enrichment of American life" did not mean thousands.[74] Klaus accepted the logic of denial at every juncture, but denial and immigration were not synonymous in his view. Shortly after his abrupt removal from the JIOA, Klaus drafted a memo summarizing his experience, in which he recalled telling the JIOA, "it was necessary to prevent the Russians from getting any German scientists . . . and that for this purpose the army should bring in military custody into the US every scientist who might be kidnapped . . . or who might voluntarily go to the Russians and assure [*sic*] that he would not leave US custody thereafter." Still, Klaus claimed, "the services never submitted until very recently any individual names of applicants although constantly urged."[75] Klaus could only conclude that the JIOA used denial as cover for its broader agenda, namely the integration of German expertise and technology into a new national security apparatus. Project Paperclip, Klaus correctly deduced, "goes much further than the reasonable requirements of military necessity."[76]

The dispute between Klaus and the JIOA extended beyond the poisonous relationship Klaus had with Wev. Nearly every JIOA staffer in a uniform savaged Klaus at the time or in retrospect, as evidenced by the antisemitic insult quoted at the beginning of the chapter, Linda Hunt's interviews with JIOA personnel still bitter at State Department opposition, and Clarence Lasby's unflattering portrayal of the "one man in the State Department" supposedly responsible for delaying immigration single-handedly. During Klaus's brief tenure on the JIOA, Wev frequently enlisted the aid of fellow officers on the committee to contain or intimidate Klaus into acquiescence. Klaus describes a tense meeting on July 12, 1946, in which several officers implied that State Department objections to immigration policy were immaterial since, one navy captain argued, "the State Department had no real place in the JIS [Joint Inquiry Staff] or the JIC or with the JCS because JCS was a military organization and the State Department position was only that of an adviser." Klaus responded that "in that case we were wasting our time" and that the JIOA seemed to reject State Department input only when the answer was unsatisfactory. After Klaus pushed for military custody yet again, Wev asserted "that we [the State Department] were imposing stricter security requirements . . . than in ordinary cases."[77] When pressed by the State Department for information on Paperclip operations and background on scientists, a reasonable demand for an agency responsible for granting visas, the JIOA rejected requests on the "grounds that the information was classified and that the State Department had no right to receive such information."[78] Invoking "national security" as a means to circumvent Klaus and his successors during deliberations over a program in which the State Department had some responsibility exemplified the militarization of national security policy.

The nadir of Klaus's year on the JIOA occurred during the February 27, 1947, meeting held at the Pentagon. The JIOA and others in the War Department arranged for an ambush of the defiant civil servant. "It soon appeared that the sole purpose of the meeting," Klaus wrote hours later, "was to obtain a commitment from the State Department by which the War Department would justify its bringing in of a list of Commerce nominees who were actually not wanted for national security but for civilian exploitation."[79] With the acquisition phase coming to a close, the JIOA expected the committee to approve a full slate of immigration applications irrespective of incomplete or derogatory files. Dean Rusk, the future secretary of state, attended the meeting as the assistant to the assistant secretary of war, Howard Petersen, along with Wev and JIOA director, Colonel Thomas Ford. Klaus recalls Ford demanding "that I agree to a certification of a list of names on his desk which I did not see. I told him that of course I could do no such thing and that certification presupposed that the department should have an opportunity to pass judgment of some kind before affixing its signature."[80] Ford said the names were classified, but that should not concern the State Department. "I pointed out that our interpretation was that we must participate in the certification of the names since the [JIOA] charter so stated," Klaus reported, reminding Ford and Wev that they could not "act without the consent . . . of the Governing committee."

The JIOA's frustration with Klaus reached its breaking point when Ford, according to Klaus, "stated in various ways that if I did not accept the paper on behalf of the State Department he would give the information to several senators who would take care of the department." Ford told Klaus he would inform his congressional contacts that Klaus wanted classified information "out of curiosity and to snipe at the program [Paperclip]."[81] Ford delivered on his threat months later, and Wev volunteered damning testimony before the Senate Appropriations Committee, in which he denounced Klaus by name.[82] These charges followed Klaus and the State Department for years, culminating in McCarthy's crusade amid the Korean War.

The JIOA's internal correspondence during Klaus's stint on the committee was not nearly as reflective and thorough as Klaus's memos, but one can discern the staffers' increasing frustration with the State Department once the JIOA realized Klaus and his colleagues refused to "facilitate" mass immigration by suspending normal procedures. JIOA staffer Captain James Horan met with two State Department officials assigned to Paperclip in May 1946 and agreed to disagree over what Truman intended by authorizing "a few" German scientists to enter the country. Horan reported that the State Department rejected the notion of bringing over one thousand "Agents of Death," as one official called them, warning Horan of the "adverse public reaction if the project was given any publicity." Horan noted the officials were "courteous and cooperative," but Horan believed his counterparts were "not inclined to support the program for the exploitation of German scientists in the US" without extensive safeguards in place.[83] Wev equated Klaus with obstruction after the State Department returned the bulk of the first batch of security dossiers unprocessed. "The reasons stated to me by Mr. Klaus for such failure," Wev wrote a navy captain anxious to secure scientists for research, "were to the effect that the records contained insufficient security information to warrant the State Department approving the issue of visa applications." Wev complained that even the "basic security requirements" Klaus outlined were impossible to satisfy. "You can readily see what the probabilities were of ever obtaining an immigration status for these scientists," Wev concluded.[84] Wev lamented to Klaus's superior, Hamilton Robinson, that he preferred to work with "responsible officers in the Department of State," who understood that "Nazism no longer should be a serious consideration from a viewpoint of national security when the far greater threat of communism is now jeopardizing the entire world." Wev bluntly told Robinson that treating Nazi affiliations as a detriment to immigration was "beating a dead Nazi horse."[85]

Wev's solution to the impasse involved controlling the flow of information from the initial military investigation in Europe to the moment a complete dossier reached the civilian agencies. Wev alerted the CIC in Germany that "there is very little possibility that the State and Justice Departments will agree to immigrate any specialist who has been classified as a potential or actual security risk." Consequently, Wev instructed investigators to consider what they include in a report given "the far-reaching effect of the Military Governor's security evaluation." A report should "only be made on the most sound premises and only when full information is available," but Wev clearly expected and received the most positive evaluation possible.[86] The results were dozens of "revised" security reports designed to satisfy meddling civilians who, Wev believed, were ultimately more concerned with protecting their agencies from embarrassing revelations than preserving internal security, let alone national security. By ascribing ulterior motives to Paperclip's critics, not without cause, admittedly, Wev seemingly vindicated the JIOA's blatant deception in handling the security dossiers.

Klaus's removal in spring 1947 signaled greater cooperation from the State Department, but it did little to expedite the backlog of applications stuck in the bureaucracies of the Departments of State and Justice. Wev and Ford despised Klaus, blaming him for a process he repeatedly told them he did not control, but the immigration proposal provoked several powerful critics who shared Klaus's fears of negative publicity and diminished internal security. The JIOA alternated between intimidation and threats against obstinate officials like Klaus and appeals to national security. In a lengthy memo addressed to Lt. General S. J. Chamberlain, the director of intelligence for the Army General Staff, Wev recounted years' worth of delays in the immigration program, this time at the hands of the Justice Department. Wev's tone was considerably more respectful when discussing Justice's opposition than it ever was with State. "During the time when delays were being encountered in the Department of State, the representatives of the Department of Justice displayed a most cooperative attitude," Wev noted, but the FBI's insistence on "a full scale investigation" inevitably delayed applications.[87]

Wev was especially alarmed by J. Edgar Hoover's strident opposition to Paperclip. Hoover reportedly told assistant attorney general Graham Morison, "I consider membership by any individual in the Nazi Party sufficient grounds for objecting to the granting of American citizenship and I consider any individual who has a long record with the Nazi Party as a threat to the internal security of the US."[88] Wev reassured Chamberlain that the JIOA was "taking great precautions to prevent the recommendation for entry of any such person," although the opposite was true. Wev assumed Chamberlain, a fellow intelligence officer, shared his worldview:

> It is difficult to understand why, in the face of the threat of communism to the security of the US, such a concentration and delaying investigation must be instigated in the cases of these specialists who are making significant and invaluable contributions to the national security of this nation through their scientific and technical efforts. In light of the situation existing in Europe today, it is conceivable that continued delay and opposition to the immigration of these scientists could result in their eventually falling into the hands of the Russians who would then gain the valuable information and ability possessed by these men. Such an eventuality could have a most serious and adverse effect on the national security of the US.[89]

Wev deemed the State Department and now, in spring 1948, the FBI's unrealistic security standards "most unsatisfactory" and a "difficult administrative burden on the military sponsoring and research agencies."[90]

Wev's passionate and thorough recounting of his ordeal with the civilian bureaucracies was soon rewarded with action. After receiving the memo, Chamberlain reached out to Hoover and asked him to listen to a presentation regarding Soviet advances in military technology resulting from the exploitation of German science. The May 11, 1948 meeting of Hoover, Chamberlain and Montee Cone of the JIOA marks a significant victory for the JIOA and the immigration program. Chamberlain clearly made an impression on the FBI director, who circulated a memo to his subordinates within hours of the meeting: "The General [Chamberlain] had a number of graphic charts of a confidential character which he showed me showing the scope of the project as well as the benefits accrued to this country from the use of these scientists over the last several years in some of the very secret scientific fields. These charts were most impressive." Chamberlain emphasized Soviet advances deriving from their own captured scientists, but assured Hoover that the Paperclippers were "probably of a better quality." After learning of the State Department's delaying tactics and Justice's own stringent procedures, Hoover "informed General Chamberlain that I would see that any name checks or field investigations in this project were given number one priority and I would like you to see that this is done and make certain that the names we now have under investigation are very promptly completed."[91] Just weeks after the meeting, Hoover expedited visas for Paperclippers in El Paso simply on the word of

Major Hamill, whose signature is found on most JIOA security forms.[92] J. Edgar Hoover was now one of Paperclip's strongest supporters.

The FBI eventually proved a valuable ally to Paperclip, but obstacles outside the intelligence community remained. As Congress debated such provisions as the Gossett Bill in 1946 and several other proposals restricting immigration to adherents of "totalitarian" ideologies, the JIOA urged its superiors to block any restrictive legislation. The JIOA notified the navy that the Gossett Bill would "exclude from citizenship . . . those preeminent specialists in nuclear physics, guided missiles, jet propulsion, infra-red rays, and the like who formerly belonged to the Nazi Party," which the JIOA noted was a prerequisite "to [a] healthy existence in Germany."[93] The JIOA wanted every senior officer and civilian appointee with influence to squelch efforts to exclude potentially valuable ex-Nazis, no matter how well intentioned the exclusion. "Their [German scientists] contribution to long range programs of research and development," the memo concluded, "would advance the scientific knowledge of the military forces and would aid materially to US national security."[94] Secretary of war Robert Patterson supported the JIOA's efforts, making similar arguments in several letters to ranking congressmen. Patterson warned a subcommittee chairperson with some responsibility for immigration that scientists forced to return to Germany "would be available for exploitation by other countries" and could potentially leak intelligence on American developments as well. "The relative position of this country in research and development would thus be doubly retarded."[95] Patterson, parroting the JIOA's standard explanation for scientists' extensive Nazi affiliations, dispelled congressional fears by claiming Paperclippers "were forced to join the Nazi party in order to survive or to be free to pursue their scientific studies."[96] Now that these consummate professionals, technocrats really, resided in the United States, their pasts were inconsequential and their loyalty ensured.

In summer 1947, after Klaus's removal from the JIOA, Bosquet Wev requested that the JCS allow him to testify before Congress and furnish the JIOA's allies with his version of events. The JCS granted permission, but insisted Wev's opinions were his own and that the testimony remain secret upon his request, informing the FBI that "disclosure of contents of this record to unauthorized persons would surely jeopardize his position in the Navy."[97] The FBI recovered the transcript in 1950 to use against Klaus and aid Hoover's campaign to discredit the State Department. Klaus was not Alger Hiss, but he and others opposed to Paperclip shared guilt by association. Wev's testimony on June 27, 1947, reveals his unfiltered opinion of Klaus and hostility toward bureaucratic obstructions hindering Paperclip. Moreover, the exchange with committee members highlights influential politicians' reflexive support for the military's national security agenda and the growing mistrust of the State Department. Wev provided all the fodder necessary to denigrate Klaus and the State Department by extension, claiming that Klaus told him he spoke for the department and openly admitted his priority was bringing in DPs, not scientists. Senator Joseph Ball, a Republican from Minnesota, asked Wev helpfully, "In other words, instead of implementing the program which you were set up to do, he [Klaus] was tearing it down?" Wev relished the opportunity to vent, calling Klaus "nasty and snotty" and listing his many affronts to the JIOA. By contrast, Wev volunteered, the former military officers occupying offices in the State Department corrected course and neutralized Klaus. "I decided that the only thing we could do now was to try to work around this man," Wev stated, complaining that the JIOA committee "called Mr. Klaus 'Gromyko' [Andrei Gromyko, Soviet ambassador to the United Nations] because everything we did, he would veto it." "In the meantime," Wev continued, "we were under constant pressure of these people [scientists] disappearing in Germany every day, five or ten men."[98] Wev wove the Soviet threat throughout his testimony, indicating that Klaus's actions indirectly helped the new enemy acquire expertise the United States lacked.

Wev encouraged his congressional allies to empathize with the plight of Paperclippers who felt betrayed by empty promises. Wev paraphrased scientists who supposedly told him, "You have been telling us we are coming over and we have disposed of our assets and sold our houses and you will not let us come. We were never good Nazis and we are just scientists."[99] The military accepted the "technocratic innocence" argument and expected the State Department to follow suit. Wev told Congress that "a scientist is a person who has very little party ideology. You give him a lab and a house for his family and a garden and he will work. Klaus will not follow that theory." This perception excused not only the scientists' Nazi affiliations and patronage, but the JIOA's perfunctory security investigations. Wev asked for and received a sympathetic forum in hopes Paperclip could proceed unhindered. "Anything you can do to help my situation would be fine," Wev concluded. Senator Ball assured Wev of his support and the committee's protection: "[I]f you get any repercussions on this thing, let us know."[100] The question remains whether Klaus's memos captured the atmosphere of the JIOA meetings more accurately than Wev's testimony, but clearly their respective accounts are diametrically opposed. The memos Klaus had written for his own files proved useful in his defense months and years later, when Wev condemned Klaus before an audience of Paperclip's supporters. Klaus was silenced, and the State Department assumed the supporting role the military required, but the negative publicity generated by Paperclip's disclosure brought additional, albeit fleeting problems.

Paperclip Goes Public

The military deliberated over when and how it should reveal Project Paperclip to the American public without provoking the same consternation the JIOA faced from the State Department. Officials reasoned that the sheer number of scientists working in several locations across the countries ensured that Paperclip's discovery was inevitable, but deciding how to spin the unusual program to the press, even sympathetic outlets, proved challenging. The military ultimately lost control of the message despite several well-crafted press releases and strategic disclosures of Paperclip's benefits in official journals. A few weeks after the initial press release, the Gallup organization conducted a poll, on January 10, 1947, the results of which indicated that the public disapproved of "importing Nazi scientists" by a margin of ten to seven.[101] Moreover, organizations as diverse as the FAS, contributors to the *Bulletin of Atomic Scientists*, veterans' groups, and the NAACP bombarded the press and the White House with indictments of the program. Albert Einstein, who was understandably appalled that the military compared him to Paperclippers in a March 1946 press release, described the incoming Germans "bacillus carries of Nazism" and lent his formidable name to the Council of Intolerance's blistering condemnation of Paperclip in December 1946.[102] Despite poll numbers and the cascade of criticism leveled against Paperclip and the military for sponsoring what some termed a potential "fifth column" inside the United States, the opposition to Paperclip dissipated within months of its exposure.[103] Most Americans, if asked, were uncomfortable with the underlying concept of Paperclip, but few felt strongly enough to encourage the Truman administration to recalculate the cost-benefit analysis of exploitation. The denial argument so effective in neutralizing internal government opposition to Paperclip within the bureaucracy seemingly resonated with the public as well. Furthermore, the military thought little of attacking vocal opponents in civil society as vigorously as it had attacked Samuel Klaus and his colleagues in the State Department.

The military decided emphasizing the "strictly limited" scale of Paperclip, at least as originally conceived, and the selectivity of the recruitment process was the best strategy for disarming critics and assuaging the public's concerns.

Paperclippers "will . . . include eminent physicists, outstanding chemists, Nobel prize winners and leaders in various research and development fields," the March 1946 press release read.[104] The War Department assured the public, "Those brought over will be carefully screened so that no active Nazis are included," but the direct appeal to national security figured more prominently than other factors: "The exploitation of these highly trained Germans will be of great value in the development of new types of weapons which were being planned by the Germans as the war ended. It will be in the national interest to use them to increase our production in many industrial fields. Due to our dwindling natural resources, it has become a strategical necessity that our nation develops substitute and synthetic materials." The release claimed, falsely, that all German knowledge belongs to the public and "not for or by single private interests." The War Department predicted accurately, however, that "[t]he value of this information to the US will probably far exceed any cash reparations."[105] After the initial outcry from the public and the accompanying poll numbers, the War Department sought to blunt the negative reaction with a Joint State-War-Navy-Commerce-Justice Department press release, announcing their intention to "conclude the procurement of German specialists under Project Paperclip on or about June 30, 1947." The actual date was five months later. The military reiterated its stringent security measures, reminding the public it worked "to prevent the infiltration of this Hemisphere by undesirable elements," an assertion the State Department disputed with just cause.[106]

Individual armed services disseminated positive stories about German scientists within months of their arrival in the United States and marshaled allies in Congress to rationalize the controversial program to a skeptical public. The army's *Intelligence Bulletin*, later the *Intelligence Review*, featured numerous stories marveling at German "wonder weapons" and the scientists responsible for the incredible breakthroughs. The Germans were "ahead in control of missiles, automatic homing devices, radar camouflage, magnetic recorders, some miniature components, metal ceramic seals, research into infrared," one article stated, but "it is clear the US will improve on this research and employ it."[107] Several articles highlighted the fantastic nature of German weapons discovered at war's end, implying that such annihilation was now within the American arsenal. The electric gun, manned glider bombs with "a suicide corps of 1,000 pilots," and the potential for guided missiles excited military planners.[108] One author cited Wernher von Braun's claims that Germany intended to build a missile, a multistage rocket called the A-10, capable of reaching the East Coast. Such advances were "proof we need to take missiles seriously."[109] With the rocket team more or less reconstituted in Texas at the time of the public disclosure, the military knew that the team's exploits would generate positive interest among the public.

The military's focus on the technological benefits of Paperclip proved effective, but absolving the scientists of responsibility for crimes still very much alive in the American imagination required more effort. Articles continued to insist that the military followed strict security protocols and recruited only scientists who would "make an 'otherwise unobtainable' contribution to American military research and development."[110] The army hoped "the American people, by their expression of friendliness and cooperation," will help Paperclip succeed.[111] Senator Harry Byrd of Virginia, a ranking member of the Senate Armed Services Committee, published an editorial in *American Magazine* repeating most of the military's arguments in favor of Paperclip, but he went even further by contextualizing the scientists' Nazi affiliations: "Were these men Nazi Party members? Yes and no, but, in general, yes. Does that lessen their value to us? It is generally conceded that scientists are interested in little except their work, and rarely in politics. But politics is often the patron of science." Byrd encouraged Americans to regard science as apolitical and scientists as technocrats indifferent to their master. They need not be revered, just employed on the right side. "In my opinion," Byrd concluded, "we are entitled to exploit these talents to our best possible advantage."[112] Whether the military bungled the public relations aspect of Paperclip proved irrelevant given the evolution of the program over nearly two decades.

The AAF compiled an official history of Paperclip within months of the end of the procurement phase in June 1947, highlighting the program's outstanding benefits, specifically financial gains and technological advances. The history revealed the bureaucratic obstacles the military continued to grapple with, naming the State Department as the worst offender. AAF historians selected memos detailing the exploits of scientists working at Wright Field under the direction of enlightened officers tasked with integrating new technology into America's arsenal. The history depicts "the State Department representative" Samuel Klaus as an obstructionist, enforcing unreasonable demands on the military and jeopardizing the safety of specific scientists as well as Paperclip's successes.[113] JIOA officers appreciated that the report contained "useful information and should be valuable ammunition when and if required to justify 'Paperclip,' " but even they recognized it would not stand up to "critical analysis."[114] Navy Captain F. R. Duborg wrote that inflated claims about "savings running into the millions" are "somewhat dangerous" given how easy it was to verify information. Duborg provided examples of "obviously incorrect" data, warning the JIOA, "The submission of such a report, unless called for, would indicate a defensive attitude and might give the impression we feel we are on unsound ground, but are trying to justify ourselves and the Paperclip program."[115] The military attempted to both win over the public after the disclosure of Germans working on US installations and preemptively neutralize critics in the bureaucracy and in Congress by emphasizing the national security argument and seemingly limitless financial benefits. Massaging the truth was hardly new to the JIOA and the armed services. Predictably, Duborg's worries about unwelcome scrutiny were all for naught.

Dean Rusk, the secretary of state from 1961 to 1969, worked for the War Department immediately after the war and dealt with the blowback from Paperclip. In a March 1947 memo to assistant secretary of war Howard Petersen, Rusk noted that "the public relations people are feeling mounting pressure on this German scientist business" and blamed the State Department, lamenting, "Our position is inherently weak because the State Department finds this whole program difficult to support."[116] Rusk occasionally fielded negative press reports on Paperclip but had seemingly had enough after learning of the Herbert and Ilse Axster disaster. In a handwritten note to an AAF colonel concerned with retaining Axster, Rusk wisely distanced himself from Paperclip's more unsavory characters: "This is one business I'm getting out of. I'd suggest it be sent to G-2 [army intelligence] for action . . . in case this guy [Axster] was a hot Nazi."[117] On balance, Rusk learned, the Axsters and Georg Rickheys of the world did little to impede Paperclip, and sympathetic journalists happily piled on the State Department as the "fly in the ointment."[118] The War Department enlisted trade journals like the *Chemical and Engineering News*, *Iron Age*, and others to publicize technical gains resulting from Paperclip while assailing "lesser officers in the State Department" and their "dilly-dallying tactics."[119] United Press correspondent Thomas F. Reynolds praised the military for transferring "German secrets," thereby devising a "reparations scheme unique in history."[120] This argument resonated with the public even if the idea of importing the Germans to the United States did not. The shriller voices of opposition originated from predictable quarters, although the military answered every charge.

Academics, especially scientists, American Jews, and the NAACP shared Samuel Klaus's distrust of the Paperclip recruits and his outrage over extending citizenship to enemy aliens while millions of DPs continued to languish in Europe. Klaus necessarily couched his objections in bureaucratic language, but independent organizations like the Society for the Prevention of World War III and the FAS issued often eloquent and scathing protests. "No matter what procedures of screening are taken we cannot trust these Germans," wrote C. Montieth Gilpin, of the Society for the Prevention of World War III, to secretary of commerce Henry Wallace. "It seems to us," Gilpin concluded, "that the desire to obtain the services of German scientists is partially influenced by the myth of German scientific superiority and achievement. This myth has been effectively shattered by the outstanding work performed by American scientists during the war."[121] The society organized in 1944 to prevent the resurgence of German militarism and to encourage thorough denazification during the occupation period. News of Paperclip clearly unnerved the group responsible for publishing *Know Your Enemy*, which depicted Germans as historically and culturally aggressive and militaristic.[122] The notion that Germans possessed innate scientific and technical abilities offended more neutral observers than the society. Prominent American scientists, understandably, had even more to say about Paperclip.

The FAS was established in 1945 by several participants in the Manhattan Project. The organization pledged to prevent nuclear war and to address the increasingly intimate relationship between scientific research and national security policy. The FAS wrote President Truman that the organization hardly discounted the need for a robust military, but using scientists "either sympathetic with Nazi aims; or, at the very least, acquiescent in them" betrayed American principles. Furthermore, the FAS continued, no amount of screening would ensure that the hundreds of Paperclippers "will not constitute a dangerous fifth column."[123] The FAS supported exploitation, as did Samuel Klaus, but declared that citizenship or similar favors "represents an affront to the people of all countries who so recently fought beside us, to the refugees whose lives were shattered by Nazism, to our unfortunate scientific colleagues of former occupied lands, and to all of those others who suffered under the yoke these men helped to forge."[124] There is no record of a response by Truman, but the military interpreted the FAS's statements as a considerable blow to the public relations campaign, especially after the press published the letter in editorial pages.[125]

Secretary of War Patterson ordered a "prompt inquiry" into the FAS and expressed his annoyance that the group did not mention the Russians seizing Germans for their own purposes.[126] At least one War Department official visited an FAS meeting, describing members as "idealistic" and unbending on their hostility to cooperating with Germans.[127] Many academics agreed with the FAS, but Charles Wilber, a scientist employed by Fordham University, wrote Truman that the FAS "is a small group of loud mouthed persons" who do not speak for most American scientists. Wilber accused the FAS of "keeping with the Communist Party Line" for urging the immediate return of German scientists.[128] Wilber was a private citizen, but several agencies in the federal government also invoked the specter of communism in the rhetorical war against Paperclip critics, in some cases years after the program had terminated the acquisition phase.

The FAS maintained a professional and respectful tone in correspondence with Truman and in a separate letter to the Navy Department, but Samuel Goudsmit, the brilliant Dutch American physicist who had led the Alsos mission into Germany, savaged Paperclip in the prestigious *Bulletin of Atomic Scientists*. Goudsmit denounced Paperclip for moral reasons, writing how sad it is "to observe that the few surviving victims of Nazism are mentally and morally starving in DP camps, while these 'Heil' shouting scientists are offered privileged positions in our country."[129] Goudsmit was most offended, however, by the notion, expressed by the military on several occasions, that German scientists were somehow more advanced than the Americans who won the war. "If we had made the serious mistake of putting our principal technical efforts upon super rockets instead of radar and nuclear research," Goudsmit reminded readers, "we too would have produced a V-2 like weapon and probably lost the war." Goudsmit, like the FAS, recognized the value of selective exploitation, but "[m]isplaced hero worship may give these men better positions and their opinions more weight than they deserve."[130] Another protest in the same issue of the bulletin asked where the State Department stood on Paperclip: "Is it possible that the State Department does not object to their becoming citizens? Does it believe that they will turn into good democrats just by receiving favors from us? And if their stay in this country really informs them, would this not be one reason more for sending them back to Germany so that they can lead their people towards democracy?" And finally, "Does this imply that permanent residence and citizenship can be bought?"[131] The resentful scientists were clearly not privy to the contentious deliberations within the JIOA. Exploitation made sense, especially when packaged as reparations or crucial for national security, but the warm embrace and rewards deeply offended a vocal segment of the population.

Academics and the scientific community were divided over Paperclip, but the cultural Left stood uniformly against it, especially as details of the immigration initiative entered the public sphere. The *New York Times* called Paperclip "unfortunate" given the United States' moral standing compared to Russia. "Whatever the drift of political events since 1945 the facts of World War II are clear."[132] *The Nation* lamented America's descent into a Cold War culture and regarded "national security" rhetoric with hostility. America's oldest left-wing publication attacked Paperclip in several articles, including one two-part piece from an intelligence officer who had been on the ground in Germany during the early acquisition phase. Morton M. Hunt, a psychologist assigned to Project LUSTY during the war, recounted his initial encounters with future Paperclippers and, most damning, his conversations with American officers completely indifferent to ethical considerations. Hunt's "The Nazis Who Live Next Door" series provided embarrassing details about selecting scientists, hiding their pasts, and settling them in the United States even as they continue to preach antisemitism and Nazi beliefs. In Hunt's profile of Werner Ditzen, a Paperclipper assigned to Wright Field, Hunt quotes Ditzen complaining that "[t]here are a few people at Wright Field who do not like us—all of us Germans. They are refugees from Germany, Jewish people, and they influence the others. You see, maybe their families were killed in Germany, or something like that. But we didn't kill their parents, or whatever it was."[133] Stories like these did little to ameliorate concerns about Paperclip. Joachim Joesten, a German émigré journalist also writing in *The Nation*, wrote a blistering tirade entitled "This Brain for Hire" just weeks after Paperclip's disclosure. "Memo to a would-be war criminal," the article began, "If you enjoy mass murder, but also treasure your skin, be a scientist, son. It's the only way nowadays, of getting away with murder. Your enemies will coddle you, and compete for you, no matter how many of their countrymen you may have killed."[134] Such venom resonated with *The Nation*'s readership, but most of the country simply registered its distaste for Paperclip and moved on to building a roaring postwar economy.

Perhaps the military lost control of the Paperclip story and misjudged the level of cooperation it would receive from the normally pliable State Department. Perhaps, too, it underestimated the hostility of the scientific elites from which the Paperclippers derived. The public opposed Paperclip, and the press coverage was generally unflattering, so why did it proceed unhindered aside from Samuel Klaus's temporary obstacles? The number of Paperclippers residing in the United States was significant, especially when one includes the number of dependents, but the Germans initially formed islands unto themselves and coexisted happily with their military benefactors. Moreover, the MIC provided all the resources, camaraderie, and purpose needed to placate the contracted scientists. The American scientific community grew in size and influence during the Second World War and maintained expansive budgets during the immediate postwar years. National security advocates encouraged permanent mobilization, and German scientists, having worked under similar circumstances in the Third Reich, recreated successful models of research and development without fearing impending defeat, capricious leaders, and a demoralizing competition for resources. The military and its Paperclippers built the MIC in concert with industry. Furthermore, scientists and intelligence officers constructed a terrifying picture of the new enemy, what Germans termed a *Feindbild.* German scientists may not have engendered much affection or even trust among Americans, but the specter of the Soviet Union proved more consequential than lingering resentment over the remnants of a vanquished foe.

CHAPTER FOUR

Their Germans

German Scientists, the Soviet Union, and the US Intelligence Community

> The alliance of German brain-power and Russian resources may well prove to be the most important outcome of the occupation of Germany.
>
> —Joint Intelligence Committee, May 1946

> To be sure, last night no one was able to sleep. This was perhaps the most exhilarating, perhaps also the most beautiful night [while we were] in Russia. Who knows? During this night, there were no differences of rank. No professors, no ministers, no military. There was only one excited, wild, big family. . . . As in the Peenemünde times, when the first tests were run. . . . There is here only one uncrowned queen, who is courted by all: the rocket!
>
> —From the diary of Irmgard Gröttrup on observing the launch of a Soviet-German missile, October 27, 1947

The Soviet Union provided both the impetus and the justification for Project Paperclip and its successor programs during the Cold War. While the State Department and some intelligence officials warned against a Nazi revival in the Western hemisphere, specifically at the hands of military research scientists, the majority of the burgeoning national security bureaucracy shifted its focus and fears to the expansive Soviet target in summer 1945. Dismantling German science and determining the extent to which the Germans shared technology with its Japanese ally was paramount, but jockeying over securing labs, equipment, and personnel contributed to the growing rift between the Allies over occupation policy and resentment over the proper apportionment of reparations. While valuable sites like Peenemünde, Nordhausen, and the network of Kaiser Wilhelm Institutes were stripped bare by whichever occupying power controlled the territory, thousands of scientists and technicians traversed the blighted country seeking food and shelter, amnesty, protection, and eventually opportunities to continue working in their specialized fields. The four Allies courted scientists inside and outside their respective zones, but the Soviet Union and the United States exploited their advantages in resources and territorial control and claimed the majority of talent. This dangerous rivalry in occupied Germany witnessed kidnappings, clandestine border crossings, and irresponsible shows of force, all of which foreshadowed a deepening mistrust over each power's intentions toward the other and anxiety over their respective military capabilities.

German scientists contributed greatly to the Soviet and American military and scientific establishments, but the Soviet Union decided to return "their Germans" in the early 1950s while the United States integrated Paperclippers comfortably into virtually every facet of the national security state. What explains the differing tracks? What insight did the returnees provide about Soviet weapons research, plans, and internal conditions and politics? Finally, how did the US intelligence community exploit and interpret information derived from German scientists peddling knowledge about the Soviet Union? In this chapter, I first detail the Soviet Union's vigorous exploitation program and the breakdown of US and Soviet agreements regarding the disposition of Germany's industrial and scientific resources. The Soviets' bold and sweeping measures to safeguard the remainder of the Nazi MIC within its grasp coincided with the disappearance of the rocket team from Europe and its reconstitution in Texas. The second topic I examine is the intelligence gleaned from German scientists, many of whom were Paperclippers; their dependents; and dozens of returnees who worked inside the Soviet Union in the late 1940s and early 1950s. German scientists painted a confusing and contradictory picture of Soviet science as either dangerously advanced or embarrassingly backward. In most cases, German scientists were anxious to secure employment, ingratiate themselves with the capitalist West, and be absolved of past crimes. Many portrayed the Soviet Union using the language of National Socialism. German scientists trapped in the eastern zones and debriefed by the CIA and FBI exhibited the same penchant for duplicity as Wernher von Braun and Walter Dornberger. In the third section, I explore the manner by which the national security state, specifically the JIOA and the armed services reliant on German expertise, exploited intelligence provided by German returnees on Soviet research and development to defend Paperclip and justify recruitment and spending. The JIOA frequently cited intelligence on Soviet advances derived from its Paperclippers to silence opposition within the bureaucracy and deflect from the limited public outcry. After contending with the State Department, the JIOA surmised that denying the Soviets' access to German military research was the most convincing argument for Paperclip. Who better to make the case for an expanded Paperclip, complete with citizenship, than the Paperclippers themselves, or, once released from Soviet employment, their counterparts from behind the Iron Curtain?

"Predatory Looting"

Major General John Medaris, the commander of the ABMA during the von Braun years, recalled that most of the national security establishment viewed the Soviets as "retarded folk who depended mainly on a few captured German scientists for their achievements, if any. And since the cream of the German planners had surrendered to the Americans, so the argument ran, there was nothing to worry about."[1] This illusion shattered with the launch of Sputnik on October 4, 1957. The testing of a Soviet atomic bomb in August 1949 shocked US leaders, mostly because analysts understood this success was less the product of imported German expertise than the combination of existing Russian scientific knowledge and American blueprints courtesy of one of the most successful intelligence operations in history.[2] The US intelligence community determined this after debriefing German scientists and POWs but did not publicize the findings.[3] US analysts feared Russia because it was a behemoth with a command economy and militarized population, as well as being cloaked in secrecy. German scientists offered rare glimpses into the Soviet MIC.

Soviet central planning, beginning with the first Five Year Plan in 1928, prioritized technological superiority, but, as Walter McDougall notes, "Rather than liberating the creativity of the people, the regime subjected the technical intelligentsia to a monopoly of patronage more constrictive than had existed under the tsars."[4] The defense sector, however, benefited greatly from political favoritism and mostly survived the devastating purges in the 1930s. Patronage and resource allocation determined success or failure in a technocratic state. While the Third Reich foolishly invested in fantastic, albeit impractical "wonder weapons," the Soviet Union and the United States choked the German enemy with a preponderance of material. The Nazi model for production mirrored the Soviet's own technocracy, but the regimes' production priorities obviously differed. Soviet rocketry and aviation suffered from a devastating war and a certain amount of prewar neglect, but the infusion of Germans certainly aided in its swift recovery at a crucial point in the early Cold War.[5]

Joseph Stalin flew into a rage in April 1945 after learning that US forces had reached Nordhausen and the V-2 production site before the Red Army, shipping nearly thirteen hundred personnel to the Western zone before vacating Thuringia: "This is absolutely intolerable," Stalin grumbled, "We defeated the Nazi armies; we occupied Berlin and Peenemünde, but the Americans got the rocket engineers. What could be more revolting and more inexcusable? How and why was this allowed to happen?"[6] While the United States eagerly anticipated demobilization, and the armed services struggled for limited resources in the initial postwar period, the Kremlin tripled the military research and development budget and initiated crash programs in nuclear power, advanced aviation, and rocket technology. Such a move, McDougall notes, "all but announced [the Kremlin's] estimate of the dangers of the postwar world."[7] The occupation of Germany involved much more than eliminating the scourge of Nazism and German militarism; it was an arena in which the two strongest superpowers competed for military supremacy. The Soviet Union systematically dismantled Germany's remaining military and industrial apparatus in part to restore its blighted economy, but the inevitable conflict with the United States raised the stakes when it came to deciding on zones of occupation and what constituted "reparations."[8] The Soviets "rehired" thousands of skilled scientists, technicians, and engineers to operate captured equipment and to provide practical expertise for reconstructed Nazi weapons systems. Furthermore, more than half of the German aviation industry (six hundred facilities) were located in the Soviet zone of occupation.[9]

Like the Anglo-American T-Forces and CIOS teams scouring Western Germany, the Red Army deployed regiments dedicated to securing scientific institutions and industrial centers as the Reich collapsed before them. These so-called "trophy brigades," composed of scientists, military officers, factory managers, and Communist Party representatives, demonstrated both precise targeting and clumsy, heavy-handed tactics. Despite explicit instructions from Major General Nikolai Petrov for the "removal, safekeeping and shipment to Moscow of all experimental aircraft and engines of all types; aviation equipment, components and all materials associated with their design and production; scientific research materials; laboratory installations; wind tunnels; instrumentation; libraries; and scientific archives," an unacceptable amount of the German infrastructure was discarded or damaged beyond repair.[10] V.

L. Sokolov, a Soviet émigré assigned to the Soviet Commissariat for the Aviation Industry at war's end, recorded his impressions concerning Soviet use and disuse of German science and technology for the American-funded Research Program on the USSR in 1955. Although catering to Cold War narratives denigrating Soviet achievement, Sokolov's recollections of the trophy brigades and the chaos of the immediate postwar period is revealing and confirmed by subsequent scholarship. The "economic disarmament" of Nazi Germany, which Sokolov aptly described as the "predatory looting of German industry," "was carried out hastily, since the Soviet leaders wished to remove as much as possible the USSR before the conclusion of an agreement on reparations with the allies."[11] Consequently, much of the dismantling was "entrusted to incompetent individuals" who mistook their mission for simple demolition. Sokolov described how Germans affiliated with the sites "thought they could fool the 'simple Ivans'" by peddling false expertise in exchange for more rations and possible employment.[12] After years of disappointing results, the Soviet Military Administration in Germany (SVAG) created "special dismantling battalions" comprising more engineers and fewer police and military officials. The engineers, a CIC report notes, oversaw the dismantlement and selected "key personnel" to accompany the machinery to the Soviet Union. "As a result of the formation of these battalions," the report concluded, "the efficiency of Soviet dismantling operations increased greatly."[13]

Interactions between Soviet and American occupation authorities began courteously enough, especially among military officers who shared a vengeful attitude toward the defeated enemy. Personnel assigned to FIAT and the Red Army's trophy brigades enjoyed a professional camaraderie far removed from macro-level concerns in Washington, DC, and Moscow. At a dinner for Allied investigators touring the IG Farben facilities, "[m]ost of the toasts and conversations were very informal as if the Russian and American representatives had known each other for a long time."[14] The Soviets certainly paid an inconceivable price in blood for whatever reparations they felt owed, but the myriad Allied agencies and special teams roving the shattered Reich often took matters into their own hands when it came to "liberating" abandoned research institutes, factories, and the remaining German personnel. Each Ally violated the spirit if not the letter of the occupation agreements. A May 1945 memo alludes to CIOS's wariness over exchanging captured technical information with Soviet units, although American members of CIOS supposedly expressed "no objection to the Russians investigating CIOS targets in the Anglo-American zone of Germany, if we [Americans] can also visit targets in the Russian zone." However, the memo continued, "Russian investigations should be separate from CIOS teams" and "have their own separate and distinct investigating units."[15] FIAT director Ralph Osborne reported that while his teams investigated freely in the British zone, "[i]t is more difficult for our investigators to exploit targets in the French zone, and almost impossible in the Russian zone." Osborne admitted that representatives from "Great Britain, France and Russia have a varying amount of difficulty in obtaining permission to investigate targets in the US zone."[16] Senior officials in OMGUS frequently learned of their representatives' transgressions in neighboring zones after the fact and contended with the political fallout from stolen booty and illicit border crossings in fall 1945 and 1946.

General Lucius Clay, the deputy military governor for OMGUS at the time, and Marshal Vasily Sokolovsky, his counterpart at SVAG, arbitrated the scientific plundering as best they could. While Stalin assailed his officers for allowing the United States to steal prized Germans before evacuating the territory, even complaining directly to President Truman, Sokolovsky found a sympathetic partner in Clay.[17] In a September 1945 letter concerning forty-nine "important specialist-scientists and men of practical experience" whom American forces supposedly took from the Soviet zone, Sokolovsky asked Clay if he would "be kind enough to issue an order for the return to the factories to the places of their former employment, of these German specialists, who are very important in the work of production of liquid fuels, so badly needed in the economy of the Soviet zone of occupation in Germany."[18] Sokolovsky avoided accusing the United States of purposefully taking personnel and equipment even when he knew it was true, preferring polite inquiries to righteous anger. The tactic forced OMGUS to respond in kind and scramble to answer the Soviets' inquiries. After Sokolovsky requested missing technical drawings and machines reportedly stored in an air raid shelter, Clay's office ordered FIAT and the US Army to "take immediate steps to locate and collect drawings and charts."[19] Clay devoted considerable time in his staff meetings to Soviet complaints and requests and frequently reprimanded officers for placing OMGUS in an "embarrassing position" by stripping factories in Soviet territory. According to the minutes from a July 27, 1946, meeting, Clay "stated it looked bad that airplane factories reported for reparations had no equipment in them between the time available for reparations and the time inspected. This matter involves loss of prestige."[20] Eventually, Clay's relationship with his Soviet counterparts deteriorated, and cataloging scientists and lab equipment seemed insignificant compared to the collapse of an uneasy peace, but the Clay-Sokolovsky correspondence highlights each administrator's challenges managing their own overzealous exploitation teams.

The Soviet pursuit of scientific personnel, equipment, and infrastructure was carefully planned and executed but ultimately less successful than US exploitation of German science and technology.[21] While US intelligence teams benefited from the exhaustive Osenberg list containing fifteen thousand names engaged in Nazi-era military research and development, the Soviets relied on a combination of draconian measures and bribery to acquire skilled personnel in bulk. SVAG administrators encouraged German engineers and scientists to remain at work regardless of their Nazi affiliations, a policy one historian noted was "reminiscent of Lenin's decision to keep on 'bourgeois specialists' after the Russian Revolution."[22] The conquering Red Army drove thousands in the Nazi scientific establishment into the interior of the country, but many scientists remained in Berlin with their institutes. In January 1946, the Soviet military mission hired several German academics to assist in producing "an exact survey of all German inventions related to the war industry 'as well as to every branch of civilian life.' "[23] The trophy brigades reportedly arrived at the locations armed with two staples of any successful recruitment, vodka and lard—"vodka for the local military commandants, whose cooperation was necessary for quickly transporting the labs," Norman Naimark writes, and "lard for the German scientists, to convince them that the Soviets were serious about taking care of them."[24] The SVAG first attempted a process of self-registration, as evidenced by an official notice disseminated in Weimar, Germany, in October 1945, ordering the following personnel to register with the mayor: "Specialists in airplanes, specialists in airplane engines, airplane weapons, specialists in establishment of airports, generals and scientific personnel who have been engaged in the airforce and in aviation." The SVAG notice offered a chance for "those who did not present themselves a final chance for registration."[25] Still stinging from the loss of the rocket team and alarmed by news of Overcast and Paperclip, the Kremlin soon abandoned protocol and polite reminders in favor of more direct measures. According to a January 1946 intelligence report, the survey of industry first facilitated "the voluntary transfer of workers . . . to Russia. It is rumored that, if they did not go voluntarily, they may be forcibly removed to the USSR."[26] Ten months later, the rumors proved truthful in the case of military research specialists residing in the Soviet zone.

The Soviet Union trained brilliant scientists in the hard sciences, but it lacked practical experience in translating blueprints for such specialized projects like guided missiles and advanced aviation designs into mass-produced weaponry. With Wernher von Braun launching V-2s in the desert, the Kremlin, according to the insider V. L. Sokolov, decided in summer 1946 "not to risk either the loss of such important German specialists or the possible failure of the attempt to establish assembly-line production without German experience, and it was finally decided that all German specialists and their families should be deported *en masse*."[27] The Soviet government instigated several deportations of skilled German labor immediately after the Reich's collapse, beginning with a few dozen Germans associated with atomic energy and the Nazi's fledgling nuclear weapons program between May and November 1945. Despite fears to the contrary, US intelligence concluded, "It appears that on the whole the Soviet atomic exploitation group captured only a small fraction of the German wartime nuclear scientists."[28] The second wave of deportations in fall 1946, better known as Operation Osoaviakhim, involved two to three thousand specialists in a variety of military research fields. The third wave, which numbered only a few dozen specialists, occurred between February 1947 and February 1948. While the intelligence community and Western media claimed the Soviets imported up to thirty thousand German scientists, not including dependents, the actual number was approximately three thousand scientists and technicians, although the number of skilled laborers was significantly higher.[29] Inflated estimates notwithstanding, Soviet aggressiveness alarmed the West, and Operation Osoaviakhim in particular sent shock waves through an already nervous intelligence community.

The name "Osoaviakhim" derives from a massive "voluntary" organization established inside the Soviet Union in the late 1920s intended to supplement labor and defense forces.[30] The name implied that German scientists would gladly volunteer to work inside the Soviet Union for fair compensation and housing for their dependents. During summer 1946, agencies like the Ministry for the Aviation Industry planned for "X-Day," set in late October, by packing equipment and materials, as well as compiling lists of personnel, some more useful than others. The Soviet order for Osoaviakhim expected a coordinated and swift detention and deportation of German scientists, engineers, and laborers who worked in the fields of aviation and rocketry. Colonel General Ivan Serov, the deputy commissar for the People's Commissariat for Internal Affairs (NKVD), under the infamous Lavrentiy Beria, issued the order without even informing the SVAG, which only found out about the deportation after it began.[31] Units of the Red Army along with the Ministry of the Interior (MVD) accompanied NKVD officials to, in Sokolov's words, make "clear the sense in which the 'voluntary' character of the contracts was to be understood."[32]

V. L. Sokolov described the night of October 21 evocatively: "While the trembling Germans, who had been roused from their beds, were signing 'voluntary' contracts presented to them by MVD men, groups of Soviet army soldiers packed the Germans' possessions into previously prepared crates which were brought up on trucks."[33] A phone call intercepted by US intelligence underscores the considerable planning and intimidation involved in the operation:

> CALLER: Last night the Russians came with rifles and fixed bayonets; and then there was a furniture van in front of the door. They took Krüger and all their things with them right away. With them were also Engelmann and his family and many other skilled workers. All of them had to go with their families.
>
> ANSWER: How ever is that possible? With furniture and everything else?
>
> CALLER: Yes, just like that, they simply had to [go].[34]

In the early 1950s, German returnees from the Soviet Union recounted their traumatic experience with Osoaviakhim during debriefings with US and West German intelligence.[35] Rolf Coermann stated that on the morning of October 22, "at 0500 hours groups of an officer with female interpreter and two or three men with machine guns appeared in every house of the selected specialists, and ordered immediate shipment of all household equipment and of the entire family to Moscow."[36] The speed and thoroughness of Osoaviakhim was impressive. Ninety-two trains filled with people, belongings, furniture, documents, and even airplanes left Berlin for the Soviet Union after a twelve-hour operation.[37]

According to the MVD, about ninety percent of the designated Germans resisted the order to sign the contracts and board the trains. Some deportees, perhaps fearing horrid conditions and poor treatment, wanted their families to remain in Germany, but the operation continued as planned. Ultimately, Asif Siddiqi estimates, "2,522 German 'specialists' boarded the trains on the morning of October 22, 1946; with their families, a total of 6,500 Germans left their home country."[38] Most of the specialists resided in Berlin and worked in the fields of aircraft and aircraft engines or electronics and electronic engineering. The most valuable specialists were those members of the rocket team who stayed behind in the Soviet zone or never made it to Bavaria with Wernher von Braun and Walter Dornberger in early summer 1945. The more prominent factories targeted by Osoaviakhim included the Junkers plant in Dessau; the

BMW plant in Stassfurt; Askania in East Berlin, which specialized in radar and automatic piloting systems; and a host of institutes housing experts in gyroscopic research, radio transmitters, and other fields crucial for aviation and guided missiles.[39]

Specialists fortunate to evade the round-up promptly migrated to the Western zones. Fortunately, for the JIOA, Osoaviakhim coincided with the debate to expand Paperclip to one thousand scientists. Ironically, the heavy-handed Soviet operation disrupted a trend in which younger German talent gravitated to the Soviet zone voluntarily because of fewer opportunities to work in the West.[40] Scientists in particular faced thorough security and political screening in the Anglo-American zones, at least initially, and the Nazi MIC infrastructure lay dormant as the Allies deliberated over the future of German science and technology. One British observer of the Soviet zone worried about alienating Germans who "had offered their services to one or another of the Western Allies and appear to feel that they had either been completely ignored or fobbed off with vague promises that show no early prospect of materializing."[41] The Soviets, contrarily, ignored denazification altogether when it came to engineers and scientists, a point the JIOA frequently cited in its bureaucratic war with the State Department. The SVAG prioritized rebuilding military research and development institutes and filled abandoned armaments factories with skilled labor.[42] Wary of Paperclip, the Soviet Union implemented its own denial program in the guise of Operation Osoaviakhim.

While Operation Osoaviakhim provided the Soviet Union with immediate results, the Kremlin impaired its future recruitment potential and supplied the United States with a valuable propaganda victory in the Cold War for German scientists. Why take such dramatic action knowing the long-term implications? Robert Murphy, the US political adviser for Germany, informed the State Department that, according to the German head of the Soviet Zonal Administration for Industry, the dismantling "was only decided upon when it became evident to the Russians that their plan of dominating the zone—politically and economically—through the SED [Socialist Unity Party] was a failure." Furthermore, Murphy reported, the German informant "thought it was very likely that Moscow would continue on this line unless Allied action prevented it."[43] With time running out in the existing agreements, the Soviet Union had to remove useful military and technical assets, both personnel and material, while they still had the chance.[44] The Soviet Union calculated that quantity would have to compensate for quality. US intelligence assessed that Soviet failure "to obtain first class scientists" required the Soviets to recruit lesser targets "with the idea ten second class scientists will be equivalent of one first rate one."[45] Exploiting its geographic proximity to central Europe and occupying prime German scientific and industrial territory, the Soviets ignored agreements to liquidate German war industries and instead resuscitated armaments factories, rehired skilled workers, and when the timing was right, uprooted the entire infrastructure.[46]

The Kremlin anticipated the protests and condemnations emanating from the Western Allies in the days following Osoaviakhim, but rather than defend its actions, Soviet officials in Germany portrayed the Soviet Union as victims of unwarranted criticism and anti-Soviet media reports.[47] Marshal Sokolovsky's cordial relationship with General Clay did not survive the blowback from Osoaviakhim. When the commander of US forces in Berlin protested the operation to Sokolovsky, the marshal shot back: "I am not asking the Americans and British at what hour of the day or night they took their technicians. Why are you so concerned about the hour at which I took mine?"[48] Robert Murphy relayed Sokolovsky's comments from a tense meeting on October 31, 1946, just one week after the Soviet operation: "The essence of his lengthy remarks is (1) the Soviet delegation feel no obligation to supply explanation or excuse for the action of the Soviet administration in transferring Germans to the USSR[;] (2) Sokolovsky resents what he insists on terming an anti-Soviet press campaign inspired by the US and UK[;] (3) the Soviet delegation promise to retaliate by means of an increased tempo of press attacks in Germany not on the basis of 'an eye for an eye but a jaw for every eye.'"[49] Sokolovsky resented the public chastising, not the collapse of working agreements concerning science and technology.

Senior Soviet scientists responsible for exploiting the Germans reportedly complained as well. Grigori Tokady, the head of the aeronautics laboratory at the Zhukovsky Air Force Academy and an expert in rocketry, told Stalin the October action had complicated recruitment for both military and civilian projects because "[t]he whole population is afraid of us."[50] Tokady fled the Soviet Union in 1948 for the West after serving as chief scientist on Stalin's Special State Commission. His intelligence confirmed, among other specifics regarding guided missile progress, that the Soviets prioritized intercontinental weapons.[51] In fall 1946, facing political pressure in its zone and reeling from an expanded Project Paperclip already responsible for securing the top-tier scientists in guided missiles, the Soviet Union gambled on Operation Osoaviakhim to fulfill its reparations demands and, perhaps more importantly, to deny the United States even more recruits. The Soviet decision to return thousands of

Germans ensnared by Osoaviakhim when their contracts ended in the early 1950s appeared to be more of a political calculation than a practical one. The United States regarded it as a potential intelligence windfall.

German Scientists behind the Iron Curtain

Anticipating a boon in captured German data, equipment, and personnel, the Soviet Union established the Scientific-Technical Council for Rocket Development under the leadership of People's Commissar Dimitri Ustinov immediately after the war.[52] Stalin, already incensed at losing the bulk of the rocket team to the United States, personally arranged for an aged artillery factory outside Moscow to house this priority project.[53] Grigori Tokady, charged with locating and securing special sites in occupied Germany, performed miracles in the field of rocketry and advanced aviation considering the disappointing number of superior German scientists within Soviet reach. Fortunately, Soviet intelligence acquired considerable knowledge of German expertise in rocketry beginning in the early 1930s.[54] As technical adviser to the SVAG, Tokady, like his American counterparts in Bavaria, specifically targeted anything associated with the V-2. "When I arrived at Peenemünde," Tokady recalled, "there was hardly a German sufficiently competent to talk about the V-2 and the other big stuff. There were many, almost all, claiming to be V-2 experts . . . [but they] talked and talked, and displayed the typical characteristics of a second-rater. . . . [N]ot only on Peenemünde, but also in all Soviet-occupied Germany, we found not a single leading V-2 expert."[55] Tokady's survey of the vaunted German missile complex confirmed three things: "(1) in the field of original ideas and rocket theories, the USSR was not behind Germany; (2) in the field of practical technology of rockets of the V-2 caliber, we were definitely behind the Germans; (3) having seen and studied Peenemünde, we came to the conclusion that there were in the USSR rocket engineers as able and gifted as elsewhere."[56] Tokady's assessment is mostly accurate, but German organizational and production methods proved more valuable than the individuals who signed contracts or were otherwise compelled to work on Soviet projects.

Peenemünde may have been stripped clean, but the Soviets found their rough equivalent to Wernher von Braun in his close associate, Helmut Gröttrup, the former assistant director of the Guidance, Control, and Telemetry Lab at Peenemünde. Gröttrup accepted Soviet patronage in summer 1945 and by mid-1946 supervised more than five thousand workers building new V-2s inside the Soviet zone.[57] The Gröttrup group, like von Braun's rocket team, saved its new client state years of expensive research and development. The circumstances surrounding Gröttrup ascending to such an important position inside the Soviet Union interested US intelligence and many of his former colleagues. A German agent in 1948 provided background to the CIC, noting Gröttrup feared for his life at the end of the war and narrowly escaped the "death train" intended to "dispose of" the rocket scientists before Allied capture. According to the CIC source, presumably another scientist, "Gröttrup did not accept the conditions offered him by the US forces, and, therefore, turned to the Soviets, who welcomed him with 'open arms' and offered him a very favorable contract with them."[58] Paperclipper Gerhard Reisig spoke at length about Gröttrup in a 1985 oral history conducted by the National Air and Space Museum. Reisig, who worked under Gröttrup at Peenemünde, called the affair "a sad story," which began with the arrest of Wernher von Braun, Klaus Riedel, and Gröttrup by the SS in August 1944. Reisig recalled, "Gröttrup was the one who was the most suspicious, because allegedly he had connections to the Communist underground organization, which was sheer nonsense."[59] Gröttrup remained on house arrest and learned from Dornberger's staff that the SS intended to shoot him in the final days of the war. When Gröttrup fled to Berlin to see his injured brother, he was supposedly contacted by both US and Soviet representatives. Reisig claimed several on von Braun's staff told Colonel Holger Toftoy, the American officer responsible for locating the rocket team, that "Gröttrup was a traitor who had delivered to the 'enemy' essential documents on the V-2" instead of going into US custody as planned, a charge Reisig called "a mean lie."[60] On the run from the SS and burned by his own coworkers, Gröttrup found a haven in Moscow and the possibility to continue his work with relative autonomy.[61]

The Gröttrup Rocket Testing Institute outside Moscow was a priority target for US intelligence, but the reliance on German informants and, later, scientists released from Soviet service, posed credibility problems for analysts responsible for assessing Soviet capabilities. The Soviets paid a price for Operation Osoaviakhim by alienating Germans considering voluntary employment in Russia and contending with negative press and diplomatic fallout, but the operation had the added benefit of complicating the US intelligence community's efforts to evaluate Soviet research and development. The Soviet deportations removed a valuable and vulnerable population of specialists from central Europe, where porous zonal boundaries enabled intelligence agencies to recruit at will. The JIOA lost potential Paperclippers, and agencies like the CIC and the newly established CIA struggled to rebuild networks providing current information concerning Soviet exploitation of German military research and development. Intelligence continued to pour in from the Soviet Union, but its reliability was undetermined. Germans living in the Soviet Union or those who corresponded with them supplied contradictory accounts about living conditions, Soviet intentions and technological advancement, Soviet-German relations, and Soviet designs on German scientists in Europe and in the United States.

As the primary intelligence unit operating in Germany, the CIC first reported on Soviet plans to reconstitute as many of the German guided missile programs the trophy brigades could find, with or without the requisite personnel. Along with the V-2, the Soviets accelerated various German surface-to-air missiles, most of which had never left the drawing board during the Third Reich. The Wasserfall, Schmetterling, and Rheintochter missiles depended on remote-control guidance, Gröttrup's specialty. Most of the engineers and technicians worked on the projects in Berlin before accompanying the labs to Moscow in October 1946.[62] The Germans contracted to the Soviets devoted the first years to rebuilding Nazi-era weapons with the remaining parts, but Gröttrup's team improved on the designs and began working on ambitious projects like the trans-Atlantic rocket.[63] Stalin personally demanded a briefing on the so-called Sänger Project, named for the German Eugen Sänger's blueprint for a piloted, winged rocket capable of skipping off the atmosphere.[64] Soviet and German personnel worked in integrated teams, at least through 1948, but this productive arrangement did not survive the next decade. A February 1948 report on a Junkers plant altered to produce the Schmetterling missile broke down the total strength of the workforce into function and nationality. Seventy-one Russian military and civilian engineers worked alongside approximately three hundred German engineers, technicians, production chiefs, and laborers, as well as one chemist and a dozen lab assistants.[65] This sort of arrangement certainly produced results, but political changes inside the Soviet Union first eliminated the integration model and eventually resulted in dismissing the Germans altogether.

The CIC conducted several operations specifically to determine Soviet recruitment methods and the conditions under which German scientists lived and worked, partly to estimate Soviet progress and partly to identify targets willing and able to work for the West as specialists or possibly agents-in-place. How the United States used this information to justify Paperclip and rationalize preparedness is discussed in the final section of this chapter. The reporting stream provides fascinating, albeit inconsistent, depictions of life behind the Iron Curtain. Although it is unlikely that each team of Germans had similar experiences, the overall trend suggests that Germans were treated well at first but that the Soviets succumbed to mistrust and a certain amount of hubris. V-2 specialists, several informants claimed, "are handled very well by the Russians" and worked "under excellent conditions," including a salary of two to three thousand Reichsmarks a month.[66] Uncommitted Germans with good reputations, at least before Osoaviakhim, received offers of five thousand Reichsmarks a month "and a small house in a settlement near Moscow," transportation for families, an automobile, and all household belongings. After surveying dozens of Germans in contact with V-2 specialists, the CIC concluded that "the Russians are making desperate attempts to contact technicians . . . throughout Germany. These people are being propositioned with high inducements in salary and in living conditions."[67] The intelligence community recorded considerable anecdotal evidence about a deliberate Soviet charm offensive targeting available scientists in every occupation zone.

Conflicting accounts from uncensored letters and refugees streaming west contradicted intelligence describing special treatment and attractive benefits for Germans volunteering to work inside Russia. In most cases, the targeted Germans had just three hours' notice before the trains departed.[68] An August 1946 situation report on the Soviet zone claimed "technical specialists" were immediately arrested as war criminals and "later released and then drafted to work. . . . Conditions under which former employees are rehired are exceptionally severe. In the contracts it is taken for granted, that sabotage or sloppy work, can draw the death penalty. The private life of these so-called 'V-2 Men' is under constant control."[69] A July 1947 report related countless tales of overcrowding, poor food rations, and scientists forced to sell furniture for extra food.[70] Another source described twenty-five to thirty German missile specialists living in "a colony of prefabricated Finnish wooden houses built by German PWs [Prisoners of War]" with "limited freedom of movement. For a time they were allowed to go to Moscow alone, but this privilege was revoked after one German allegedly went to the US legation."[71] Discerning the truth from biased informants, whether from dedicated anticommunists or self-professed Russophiles, complicated the intelligence community's search for clarity.[72]

Most intelligence reports on Soviet exploitation of German technology and refurbished factories indicate that while skilled Germans received attractive offers and some independence, unskilled laborers suffered the full brunt of Soviet coercion. Days after Osoaviakhim, German refugees stated that the Soviets used approximately sixty to seventy thousand internees to work in "the sub-terranian factories" where they were "quartered in the barracks of a former concentration camp [Dora]."[73] In March and April 1947 the Soviets commenced Aktion AU to "draft 52,000 persons into labor gangs for the purpose of dismantling industries in Thuringia and mine Uranium."[74] The Dora camp, analysts believed, was likely in use for "armament purposes."[75] Technical help worked longer than regular labor, but they reportedly received "food and clothes bonuses, regardless of party affiliations."[76] One refugee who fled the Soviet zone recounted the aftermath of a failed escape at a "concentration camp," presumably Dora. The source joined "[w]omen, old men, children, war invalids" on trucks headed to Nordhausen, where his personal belongings and identification papers were confiscated. After the Soviets dynamited the tunnels, the source witnessed them calling names out. "These were the men who had escaped from the Nordhausen camp," he stated. "What one did to these men is too horrible to describe. Around 100 men armed with black jacks, gun stocks, sticks, or any other instrument they could get their hands on threw themselves on the 57 men. In less than five minutes these men lay unconscious, covered with blood on the ground."[77] Intelligence reporting indicates the Soviets offered generous recruitment packages to elite scientists while subjecting unskilled German labor to brutal measures. Once transferred to the Soviet Union, the communities of German specialists came to respect their new patrons, but also fear the consequences of failure.

Each Allied power discovered that using German scientists to recruit their colleagues achieved better results than intimidation, radio broadcasts, or lofty promises from foreigners arriving unexpectedly at the doorstep. The Soviets dispatched dozens of German agents, usually scientists, who volunteered to track down colleagues and urge them to work in either the Soviet occupation zone or the Soviet Union. Perhaps wary of defections, the Soviets asked German specialists to write letters to contacts in the Western zones. The CIC acquired several examples, noting that the "letters paint a pleasing picture of the living conditions there and the treatment offered by the Soviets."[78] Heinrich Kaefer, an electrical engineer, received a letter from a German colleague that was typical of hundreds sent to Germans in the Soviet zone: "In the letter, [Professor Ferdinand] Ruhle stated how good conditions were in the Russian zone and wanted me to join him there. He said in the letter that he was well treated by the Russian[s], and that if I joined him there, I would do the same type of research work I did during the war. This was V-2 rocket and jet propulsion experimentation." Another electrical engineer was approached in person by an unknown German offering a similar contract.[79] US intelligence officers who were invested in Paperclip and an expedited security evaluation process for the "ardent Nazis" on the target list claimed that German scientists would continue to accept Soviet offers so long as they believed their talents were wasted in the Western zones. Citing the intercepted letters, the CIC noted that the Soviets promised "work and conditions similar to that which was offered them in Germany before the capitulation."[80] Yet, these same Germans enticed to work for their former enemy were unceremoniously dismissed and sent back to Germany less than five years later.

The Soviets promised resources and autonomy to their German recruits in 1945 and 1946, but this generosity devolved into neglect and paranoia by the end of the decade. While Wernher von Braun demanded and received nearly everything he asked from his American patrons, even directing the army's research and development agenda, Helmut Gröttrup's team struggled to survive the Soviet Union's byzantine political environment in the twilight of the Stalin era. According to Gröttrup's mother, who received mail from her son through an intermediary, Gröttrup spent three months in a "concentration camp" in 1948 for "his failure to keep to the schedule set by the Russians. He has now suffered a nervous breakdown and is recuperating with his family." CIC agents interviewing Gröttrup's mother related rumors of dissension within the team at the Rocket Testing Institute in Moscow: "Numerous quarrels are breaking out among the scientists due to strained nerves and differing political opinions which are kept from flaring openly through the strenuous intervention of the Russians. Furthermore, many of the dependents of the scientists have become so sick of the drastically simple life forced on them that they wanted to leave their men and depart from Moscow."[81] Gröttrup wrote favorably about the Soviet Union at first, even traveling to different cities to visit family, but "[s]ince the beginning of 1948 . . . conditions in general, as well as freedom of movement, were greatly restricted." In one of Gröttrup's last letters, "he plainly indicated that if it were not for the safety of his family he would have attempted an escape from Russia some time ago."[82] It is difficult to verify the accuracy of such reporting, especially given the source, but Gröttrup was just one of many scientists working in the Soviet Union who shared similar stories with their families. Still, Germans in Russia consistently produced results, even under duress. The Gröttrup team successfully launched V-2s in autumn 1947, and in 1949 the team built the R-14, which had a range of eighteen hundred miles and could potentially carry an atomic weapon.[83]

The decision to release hundreds of German scientists from Soviet service seems inexplicable at first, especially in comparison to the United States' obsession with denying the Soviet Union specialists in weapons research, but the Germans returning from the Soviet Union indicated that the regime had isolated and segregated them from the expanding Soviet MIC in the early 1950s. After assisting the Soviets with building Nazi-era weapons, many Germans reported that the Soviets assigned their own scientists to develop future iterations of the original German designs. The Germans acted as consultants and trainers, often performing the most rudimentary tasks regardless of their expertise. As Peter Lerten, a German scientist "abducted by the Soviets," described, "The Russian purpose in the first five years was simply to pump us dry of everything we knew in our field. . . . At that time they were well behind Western science practically everywhere. But they were determined to catch up and by and large they did."[84] Returnees questioned by Western intelligence agencies did not even know if their work resulted in an actual weapon.[85] The wife of Wilhelm Menke thought the Soviets and Germans engaged in a friendly rivalry, "but it later became increasingly more competitive" until 1951, when the Soviets "had largely pushed the Germans aside."[86] West German intelligence confirmed that from October 1951, rocket specialists "were deprived of any right to control . . . further development. From that time onward the Germans were occupied with the construction of general special instruments exclusively."[87] Once the Soviets decided to return German specialists, even allowing them to live and work in the West, the Germans were stripped of responsibility and kept ignorant of Soviet technical developments for at least a year prior to repatriation. Consequently, the Soviets expected Western intelligence agencies to acquire outdated intelligence.[88] The Soviet tactic worked in that the returnees related considerable historical information on the period 1945 to 1947 but minimal actionable intelligence after that.[89] Those returnees preferring to settle in East Germany were, Dolores Augustine writes, "welcomed like a 'nobility without titles' in the Workers' and Peasants' state" and had "brilliant careers in East German industry," courtesy of lavish state sponsorship.[90]

The seclusion and eventual dismissal of the very scientists the Soviets devoted so much effort into securing must be interpreted in the context of Soviet politics in the first decade of the Cold War. Asif Siddiqi, an authority on the history of Soviet science and technology, placed the German specialists at the center of the Soviets' "articulation of a Cold War national identity" characterized by a "resurgence of postwar nationalism" and the "culture of

German scientists and others at Friedland, a way station on the border of the Soviet and American occupation zones, February 1958. After exploiting German scientists for nearly a decade after World War II, the Soviet Union separated the Germans from sensitive weapons projects and returned them to both East and West Germany in the fifties. *Source:* Bundesarchiv B 145 Bild-F005116-0001, photograph by Egon Steiner, CC-BY-SA 3.0.

extreme secrecy" fundamental to Soviet science.[91] Eventually, the United States integrated Germans into every facet of the MIC, granting them access to information and empowering them to make programmatic decisions. The Soviets, on the other hand, "worked within a parallel structure that served as a comparative model for the Soviets' own work."[92] Soviet engineers exploited German expertise but jealously guarded their own work, even when such arrangements hindered progress. Senior Soviet leaders, specifically followers of Stalin's reported successor Andrei Zhadanov, launched a thorough campaign to root out "any and all vestiges of 'Western' influence in Soviet culture." The campaign continued after Zhadanov's death in 1948.[93] Initially courted and privileged, the community of German scientists found themselves engaged in meaningless work a few years later. As the CIC reports confirmed, teams of Germans succumbed to infighting and alcohol abuse before being sacked and deported beginning in 1951. Siddiqi reveals that the Soviet impulse to craft a new postwar identity required repudiating German expertise. Reframing the Germans as "less useful" meant more in political terms than mining their scientific expertise indefinitely. "In some sense," Siddiqi concludes, "both the arrival and the departure of the Germans were driven by Cold War concerns."[94] US national security advocates discovered that information gleaned from returnees' proved useful for promoting increased defense spending and an aggressive stance toward the Soviet Union. German scientists, both the Paperclippers and the returnees from the Soviet Union, recognized an opportunity to promote themselves as well.

Discerning Soviet Intentions and Capabilities

The race for German scientists originated before the collapse of the Third Reich and the deterioration of relations between the United States and the Soviet Union, but the onset of the Cold War played into the hands of the JIOA and other Paperclip promoters working in every corner of the national security bureaucracy. Denying the Soviets potential expertise in military research and development superseded lingering concerns over German scientists' Nazi affiliations, revanchist tendencies, or criminal activities, past or ongoing. US officials reading the intelligence reports emanating from the CIC and later, the returnees, concluded that the plight of displaced German scientists in central Europe made the scientists susceptible to Soviet recruiters acting on behalf of a regime indifferent to the same moral and security issues complicating Paperclip. For the United States to equivocate over Paperclip, specifically the immigration question, while the Soviets initiated mass deportations like Operation Osoaviakhim seemingly endangered national security. Intelligence derived from German scientists contributed to US national security advocates' increasingly menacing perception of the Soviet Union. Aware of their precarious position as contractual employees working inside the United States, or returnees lobbying for opportunities in the West, German scientists communicated harrowing stories, questionable estimates, and anti-Soviet propaganda to an intelligence community eager to confirm its own self-serving conclusions.

Discerning Soviet motives in Europe understandably consumed the small, overburdened intelligence community, but most estimates assumed the Soviet Union would not precipitate a military conflict with the West for several years after World War II.[95] Many agents and operators in Germany initially interpreted Soviet actions as those of a ravaged ally anxious to extinguish the Nazi threat and bolster its exhausted armed forces. Gerold Robinson, the chief of the State Department's Research and Intelligence Department admitted in December 1945 that "[t]he problem of Russian capabilities and intentions is so complex, and the unknowns are so numerous, that it is impossible to grasp the situation fully and describe it in a set of coherent and well-established conclusions."[96] In early 1946 the emerging national security bureaucracy found the explanatory model it sought in George F. Kennan's eight-thousand-word "Long Telegram," outlining the Soviet Union's ideological worldview and subsequent postwar agenda. Stationed in Moscow and the State Department's most seasoned observer of Soviet politics, Kennan advocated patience and firmness in dealing with an intractable foe. The Soviets' unique brand of communism, Kennan wrote, was informed by Stalin's fears of "capitalist encirclement." Accordingly, Kennan continued, the Soviets believed, "Everything must be done to advance relative strength of USSR as factor in international society. Conversely, no opportunity most be missed to reduce strength and influence, collectively as well as individually, of capitalist powers."[97] Kennan's unambiguous analysis and confidence reverberated throughout the bureaucracy and encouraged similar policy papers predicting a long and costly struggle with the Soviet Union.[98]

In September 1946, Clark Clifford, Truman's closest political adviser, presented his assessment of relations with the Soviet Union in a document citing distressing reports from Germany: "Soviet espionage activity has flourished, German scientists have been kidnapped, former pilots now working in the US zone have been enticed into the Soviet zone, Soviet agents have illegally entered the American zone for the purpose of collecting documents on German atomic research, and German jet propulsion experts have been recruited through German intermediaries for service with the Soviets." Fearing a "total" war with the Soviet Union, Clifford argued that "the US must be prepared to wage atomic and biological warfare" and invest more in "highly effective technical weapons."[99] The Clifford Report, which ignored contradictory evidence suggesting the Soviets were militarily weak and actually cautious, coincided with Truman's decision to authorize Paperclip and even expand its scope. The War Department sold Paperclip as both a denial program and a boon for the US military's research and development initiatives.[100]

The intelligence community rarely spoke with one voice, but organizations like the JIC, the Central Intelligence Group (CIG) and later the CIA tempered fears of Soviet intentions with dispassionate analysis of their limited capabilities between 1945 and 1947.[101] The Soviet Union acknowledged its weaknesses, analysts determined, and needed "to avoid . . . conflict for an indefinite period" while engaging in an "intensive effort" to acquire "special weapons" like "guided missiles and the atomic bomb."[102] On October 31, 1946, just a week after Operation Osoaviakhim, CIG analysts produced a prescient and restrained assessment of Soviet military advancements based on "the current estimate of existing Soviet scientific and industrial capabilities, taking into account the past performance of Soviet and of Soviet-controlled German scientists and technicians, and our own past experiences, and estimates of our own capabilities for future development and production." The report predicted the Soviet Union would produce V-1- and V-2-caliber weapons by 1950, but "[t]he possibility that the German A-9, A-10, and associated missiles may be developed to an effective range of 3,000 miles within the next ten years is considered remote."[103] Nonetheless, combining legions of German specialists, many of whom enjoyed a near mythic reputation among some military officers, with Soviet centralization and natural resources represented a grievous threat to future national security.

The US Army's *Intelligence Review* distilled various reporting streams in an attempt to gain insight into the German influence on the Soviet MIC. One of the more measured reports contained both accurate numbers of German scientists and realistic predictions of their impact. "The Germans now in the USSR," an August 1948 article read, "are experts in taking lab ideas and translating them into actualities on a mass production basis." The US Army's experience with von Braun's rocket team confirmed this, but the article predicted that the Soviet Union would likely squander German talent since much of the "supervisory staffs are operated by people who are not technically qualified . . . and must refer everything to Moscow for a decision."[104] Throughout the Cold War, but particularly in its nascent stage before the Korean War and NSC-68, the intelligence community wavered between rational forecasts of Soviet intentions and capabilities and frantic overreactions based on an incomplete picture.

The national security bureaucracy existed in part to serve itself, and intelligence reporting, accurate or not, became as commodified as any other resource. On March 5, 1948, General Lucius Clay cabled the JCS with an unsettling account of his meetings with Soviet officers in Germany. After first reiterating his position that the Soviets were unlikely to provoke war, Clay wrote, "Within the last few weeks, I have felt a subtle change in Soviet attitude which I cannot define but which now gives me a feeling that it may come with dramatic suddenness. I cannot support this change in my own thinking with any data or outward evidence in relationships other than to describe it as a feeling of a new tenseness in every Soviet individual with whom we have official relations."[105] The cable prompted the Spring Crisis of 1948 and played directly into the hands of the national security agenda. The CIA rejected the threat of an imminent attack, agreeing with Clay that nothing concrete supported his hunch, but acknowledged that "the USSR might resort to direct military action in 1948, particularly if the Kremlin should interpret some US move, or series of moves, as indicating an intention to attack the USSR or its satellites."[106] In other words, the United States might provoke the crisis predicted in Clay's cable by overreacting to Clay's cable. The armed services seized on the communication to lobby for massive budget increases. The air force's 1949 budget, which was altered during the crisis, doubled the 1948 figure. The navy had similar success.[107] James Webb, director of the Bureau of the Budget, accused the military of scaring the country "into believing that anyone who wouldn't go along with these plans would be responsible for a catastrophe."[108] Clay revisited the crisis in 1950, writing in his memoirs that he never intended to draft an "alarmist report" contradicting intelligence reporting. Clay felt vindicated, however, after the Soviets initiated the Berlin blockade in June 1948.[109] The 1948 "war scare" resonated with a bureaucracy prone to interpret Soviet actions as pretexts for invasion. The increased tension further justified safeguarding German scientists in Europe and retaining the ones working inside the United States.

The bureaucracy formed by the NSA of 1947 responded to the Spring Crisis and Berlin blockade with an impressive array of contingency plans. On November 18, 1948, the National Security Resources Board (NSRB) convened an extraordinary meeting with the JIOA, the Commerce Department's Office of Technical Services (OTS), and the chief of the Scientific Research Group for OMGUS to simplify the procedures for obtaining "the benefit of [German scientists'] services to this country and . . . prevent the Soviet Union from establishing the benefit of their services." Embroiled in the "Berlin crisis" and wary from the previous few months, the participants devoted considerable time to a "plan ready for immediate implementation for the removal of key scientists and other personnel, not only from Germany, but from all of Western Europe in the event of a Russian sweep to the sea." The NSRB planned for the evacuation of Dutch, Belgian, French, and British scientists in addition to all remaining Germans. The JIOA seized the moment, urging "cutting through red tape in the State and Justice Departments. For example, changes in emigration [*sic*] laws to induce foreign scientists to come to this country." The NSRB agreed to a scorched earth policy in the event of war, advising the destruction of "cyclotrons and other research lab equipment" to prevent Soviet capture. While the meeting was held in a crisis atmosphere, most of the attendees supported the simpler policy of offering scientists better contracts and ignoring derogatory security information when vetting candidates.[110]

Although there was no Paperclip for Japan, the Spring Crisis prompted US measures in the Pacific as well. Japanese scientists were in no danger of being overrun or kidnapped by Soviet forces, but US military planners recognized the potential Japan represented for staging US forces during a sustained conflict. "There can be little question that the establishment of well-equipped and well-staffed research labs in Japan would be an important asset in the conduct of theater level research in event of war," a SCAP memo from January 1949 declared. Citing the precedent of British labs supporting European operations during World War II, SCAP advocated erecting and resourcing in Japan "a number of labs staffed by competent technicians and equipped with basic scientific apparatus capable of wartime augmentation or adaptation for specialized military technological purposes."[111] General Douglas MacArthur considered rebuilding Japan's scientific infrastructure crucial for his mission to both stabilize the Japanese economy and provide for its defense. SCAP conducted a thorough survey of research labs and identified candidates to receive technical advice, equipment, access to previously censored research, and, of course, funding.[112] An estimated seventy thousand Japanese scientists "represent the sole remaining major potential able to turn the products of industry to modern war uses," a SCAP report determined, and though the point of the occupation was to prevent any resurgence in militarism, the crisis of 1948 required incorporating this personnel into contingency plans.

The JIOA never lacked ammunition for its campaign to broaden the scope of Paperclip and justify an expedited immigration program thanks to Soviet aggressiveness and a steady stream of reporting that originated with frightened and impoverished scientists. The specter of a mass kidnapping of German weapons specialists had tormented US authorities since the first weeks of the occupation. Operation Osoaviakhim seemingly validated these concerns. The CIC commenced Operations Hell, Harass, and Mesa in early 1946 to surveil "the activities of German scientists and technicians living in the US zone of occupation in Germany, and the efforts of foreign governments to recruit them for employment in the other zones of Germany and in foreign countries."[113] The CIC operations first tracked Soviet scientists, specifically nuclear physicists, who entered the US zone and contacted their German counterparts. In its justification for Mesa, the CIC wrote that it "feared that there may be an attempt on the part of the Russians to spirit away these German specialists and use them for secret research in Russia."[114] Mesa existed to both monitor the Soviets and protect the German scientists. Operation Hell was a "locally run" operation providing around-the-clock protection of a German nuclear physicist who was a likely target for "hi-jacking . . . by Russian agents."[115]

The notion that German scientists were vulnerable to Soviet machinations, specifically "running nets of Germans recruiting scientists [and] technicians from the US and British zones," served to expedite Paperclip.[116] Colonel Donald Putt at Wright Field cited the "continuous and increasingly alarming reports" from "extremely reliable resources" that simply detaining scientists in temporary camps invited disaster. Putt wrote his AAF superiors several times in spring 1946 relaying the intelligence garnered from the CIC operations, warning, "The American zone is literally crawling with French and Russian agents whose work has become rather fruitful and facilitated by the sorry fact that German scientists have received no clear cut, positive offers from this country."[117] Putt "requested that emergency action to procure additional Project Paperclip personnel be initiated immediately . . . if we are to divert the services of valuable scientists from France and Russia to the US."[118] The JIOA understood that the Soviets recruited primarily through bribery and lucrative contracts, not kidnapping, but lurid details surrounding these Soviet operations in the US zone roused the bureaucracy into action.

Preventing interzonal crossings was hopeless, and surveilling hundreds, if not thousands of scientists with limited manpower proved impossible. Army intelligence scrambled to prevent Soviet recruiters and disrupt what many operatives assumed were attempted kidnappings of prominent Germans. The Berlin criminal police informed the CIC that 337 people disappeared from the city in June 1946, 245 from the Soviet sector.[119] The CIC acquired some insight into Soviet methods, many of which were clumsy and amateurish. Soviet intelligence abducted targets at train stations, shoved Germans into running cars, and lured scientists into hotel rooms with attractive women. The CIC described one such "honey trap" scenario in explicit detail after interrogating the woman hired by Soviet intelligence. A German Jewish refugee who fled Germany in 1939, the woman married a Soviet soldier who was declared MIA during the war. After moving to Vienna to work as a nurse, a Soviet officer named Orlov admired her beauty and offered her five thousand Austrian schillings to pretend to be in distress and then proposition the targeted scientist, luring him to a waiting vehicle. The attempt failed, but the CIC used the account to underscore scientists' vulnerability in any of the occupation zones.[120]

Unsolicited letters from concerned Germans also contributed to US fears. An aviation engineer named Heinz Hellmond wrote General Charles Saltzman, the assistant secretary of state for occupied areas, that all his fellow experts needed rescuing from the Soviet zone "so that they won't have to be exposed to these Russian gangsters."[121] The CIC was impressed with the lengths to which the Soviets went to deceive their targets. Operation Osoaviakhim frightened Germans, understandably, so Soviet couriers, usually Germans employed by Soviet intelligence, personally delivered offers, helped recruits memorize instructions on crossing the border, and assured scientists of the excellent conditions. The CIC reported, "Every courier is . . . instructed to state that he, himself, made the trip and is to erase all doubt as to the possible return of the scientists by bringing his own case to the foreground with an 'I came back, didn't I?' "[122] The intelligence community portrayed the Soviets as simultaneously reckless kidnappers and accomplished seducers. In either case, the perception remained that the West was losing the war for German scientists.

Taking the various reporting streams into consideration, organizations invested in exploiting German science encouraged more aggressive measures to recruit and ultimately retain German scientists. A FIAT memo to US diplomat William Culbertson exemplifies the argument: "The voluntary movement of German scientific personnel into the USSR or Russian controlled territory is not difficult of explanation. At present, according to reliable reports, the working and living conditions provided are satisfactory, food rations are generous and salaries are very high. The importance of these favorable conditions increase as the present winter progresses. Of at least equal importance is the consideration that only in the Russian Zone of occupied Germany are there opportunities for German scientists to carry on research and other scientific work."[123] Every vested party in Paperclip, from the JIOA to the contracted Germans and future recruits, argued that the United States had to do much more to mitigate Soviet inducements. US officers boasted that Paperclip secured excellent scientists despite "the fact that salaries paid in Russia are often five times greater than the salaries paid to the scientists in the US," but the Soviets still enjoyed greater success in the occupied zones.[124] Timing was everything, the JIOA warned, noting that key figures like Gustav Hertz, renowned nuclear physicist, and Nobel Prize–winner Otto Warburg had voluntarily left for the Soviet Union.[125] If scientists escaped Europe altogether for friendlier environs like Argentina, so much the better. Waldemar Otterbeck, a senior Messerschmitt aircraft engineer, walked into the US embassy in July 1948 offering his services in what he knew to be a seller's market. Otterbeck claimed he was being courted by both the Soviets and the British but preferred US patronage.[126]

Shortly after the war, the US military enlisted sympathetic press to highlight the perceived gap in pay and opportunity between the West and the Soviet Union and to embarrass Paperclip's critics. In an article entitled "German Scientists Find Reds More Co-operative than Allies," the *Washington Daily News* quoted a "former State Department official" who claimed the Soviets gave scientists "every encouragement," while those in the Western zones "are handicapped by confusion and lack of funds and facilities."[127] Other news stories circulated during the immigration debate highlighted the risk inherent in returning scientists to Germany. "Sure we coddle them outrageously," one defense official told a reporter. "We have no choice. We can't afford to let them go back, knowing as much as they do."[128] The JIOA's political allies in the conflict with the State Department took to the airwaves. Senator Styles Bridges, appearing on *Meet the Press* in July 1947, noted he and his colleagues were "concerned . . . that America hasn't secured her share of the German scientists, while Russia has picked up a larger number and is utilizing them to the full extent." Bridges specified jet propulsion, missiles, and germ warfare.[129] If the United States was to compete and win, Bridges intimated, the nation had to forgive and forget the scientists' Nazi pasts.

The JIOA's motivations for depicting German scientists as miserable, underemployed, and hopelessly exposed to Soviet influence were fairly transparent to Samuel Klaus, who recommended that the United States implement its own version of Operation Osoaviakhim to remove scientists from harm's way rather than invite an indeterminate number of enemy aliens into the country. The JIOA always considered permanent employment preferable to a "catch and release" strategy. Consequently, the JIOA determined, the United States' own denazification regime hindered Paperclip as much as Soviet recruitment and kidnapping did. An August 19, 1946, memo sent to JIOA chairman Bosquet Wev related the familiar stories of generous Soviet offers to prominent Germans and concluded with this warning: "The fact that both the Russians and French governments ignore the German Scientists' previous political affiliations has a demoralizing effect upon those whom the US consider possible candidates for exploitation."[130] Army intelligence issued its own memo just weeks after the spring 1948 "war scare" on the need to exploit German scientists outside Europe. Citing the threat of invasion and the diminishing morale of scientists stuck in limbo, the army blamed delays on useless questions like "the degree of participation in Nazi party." Echoing the JIOA, the army argued that most valuable Germans were party members, albeit "nominal Nazis," and that "danger from Nazis is infinitesimal as compared to benefits for national interest during military exploitation and later employment in civil industry."[131] Contrarily, according to the security standards agreed to by the JIOA and the FBI dated the same week as the memo absolving Nazi Party members, "Communistic affiliations or inclinations are regarded as a basis for an unfavorable security evaluation."[132] The FBI personnel investigating Paperclippers and the returnees in the early 1950s consistently probed subjects for even the slightest possibility of Communist influence, but Nazism rarely counted against a potential recruit.

Despite Soviet efforts to isolate German scientists and compartmentalize their research projects years before releasing the scientists for possible employment in the West, the returnees provided valuable intelligence on the MIC inside the Soviet Union. The Germans debriefed by the CIA and the FBI lacked specifics on Soviet scientific and technical achievements in the early 1950s, but the interviews garnered targeting information for future intelligence operations, specifically routes for spy planes and other collection operations.[133] The British ran a parallel program called Dragon Return, dedicated to studying Soviet advances by running German agents and debriefing returnees working in aeronautics and guided missile research.[134] British intelligence reached US analysts through existing sharing arrangements. The returnees related inaccurate and biased assessments of Soviet progress, but offered rare insight into daily life for foreign experts in the Soviet Union. The CIA issued a report entitled "Scientific Progress in the USSR" in October 1953, derived exclusively from returnees who managed to both deride Soviet progress and elevate their own importance. "As soon as the Soviets took over a project," the report read, "it became evident that the work progressed more slowly."[135] Günther Bock, a specialist in aerodynamics who worked in the Soviet Union from 1946 to 1954, concluded that Germans had helped the Soviet Union achieve parity with the West in aviation and "other technical fields such as the mining industry, the heavy machine industry and similar branches."[136] Aerospace medicine pioneer

Hubertus Strughold repeated rumors of "undetectable U-Boats" and "secret weapons in Soviet Russia, such as V-3, giant rockets and so on."[137] German scientists were asked about their opinions concerning Soviet intentions and capabilities in Europe, but analysts simply reported the information without assessing it. Most returnees and Paperclippers interviewed by the FBI and the CIA were evaluated either as counter-intelligence threats or opportunities, not as experts on Soviet science.

German scientists happily reported stories of maltreatment and Soviet backwardness, although not all were disparaging of life inside the Soviet Union. Jürgen Karl Heinrich Rottgardt claimed, "Every minute in the USSR was like living in hell. We were living under the most severe restrictions and felt always spied upon, not only by the Soviet officials but also by our own colleagues. . . . Hardly anyone returned from this ordeal sound in body and mind. I have suffered severe heart attacks and most of my fellow workers are suffering from depressions and nervous disorders."[138] Ursula Schaefer, the wife of a scientist, claimed "My experiences in the Soviet Union will last me a lifetime, and I am going to do everything in my power to prevent this system from spreading over the rest of Europe."[139] Engineer Kurt Thöm considered it ironic that the Soviet Union is known as a "worker's paradise" given the horrible conditions, noting that he will "never again support as a voter a Marxist party after I have seen the conditions in the Soviet Union."[140] Herbert Axster, sensing his waning relevance as a Paperclipper and considered an extraneous colleague of Wernher von Braun, portrayed himself as a dedicated Cold Warrior in a 1949 interview with the FBI, volunteering that "he despised the Russians and believed the US should enact a more stringent safeguard against communism inasmuch as Russia had gained so much power in Germany." Axster told the interviewing agent about his experience with Russians during the war, describing them as "filthy and cruel." He claimed Russians soldiers grazed on grass when out of food. The FBI thought Axster "made the point to show that the Russians would be extremely difficult to conquer inasmuch as they were more like animals in their endurance than other peoples."[141] Georg Knausenberger assumed war was "inevitable," and the West waited to strike at its own peril given "Russia's continued technological advance."[142] Returnee Ernst Schaaf "found my hatred against the Soviet regime confirmed. . . . I find this system a menace to mankind; I am ready to combat it whenever I find it."[143] Similarly, Eitel Spiegel apologized for working to "the best of my ability" for the Soviets. "Now, I wish at last to serve the forces which are pitted against the Eastern colossus."[144] The intelligence community dutifully reported the Germans' anti-Soviet rants and seemed pleased to confirm suspicions, but such sentiments yielded no actionable intelligence.

The "Red scare" mentality of the early 1950s encouraged the FBI to closely examine, and in some cases reexamine, Paperclippers' Soviet connections and disqualify recent returnees from consideration as employees of US-based industries. A 1951 FBI memo reveals that many in the intelligence community distrusted the returnees and believed that "60 percent of the German scientists being returned from Russia and Russian occupied territory were sympathetic with the communist ideology." Consequently, "there was no plan to bring German scientists previously employed by Russia to the US for employment." After helping establish a German Scientists Interrogation Program to manage the influx of data from returnees, the FBI requested that the intelligence community provide "information concerning German scientists in the US indicating (1) they were sympathetic with Soviet aims; (2) that the Soviets were aware of the employment and work being done by the German scientists, and (3) that the Soviets had indicated an interest in the scientists in the US."[145] Nearly every Paperclip scientist interviewed was evaluated as a possible counter-intelligence asset, or at least as someone trustworthy enough to report anything suspicious to the FBI. Hermann Kurzweg, an expert in supersonic aerodynamics, was deemed "one who would represent attractive bait for Russian spies. He has the intelligence and capacity to function effectively in the role of a double agent."[146] Walter Dornberger also impressed the FBI as "a very intelligent man with a broad understanding of Soviet techniques . . . and could carry on satisfactorily in the role of a double agent."[147] Less prominent scientists promised to report whether they or their relatives in Europe "had been contacted by Russian officials for espionage or sabotage purposes."[148] Predictably, several scientists accused each other of being Russian spies, prompting the FBI to expend resources chasing dubious leads and adjudicating internal squabbles.[149] German scientists generated considerable intelligence on the Soviet MIC, even if historical in nature. Yet, evaluating the intelligence involved parsing the Germans' motivations, access, and reliability, which often required more work than the information was worth.

Harry Rositzke, an OSS and later a CIA officer for twenty-five years, recalled the Cold War environment in the late 1940s: "[T]here was no longer any uncertainty. . . . The Soviet Union was the enemy, and the 'Soviet target' our intelligence mission. We were professionally and emotionally committed to a single purpose. We felt ourselves as much a part of the American crusade against Stalin as we had against Hitler."[150] The national security state shared this worldview and overcame opposition to Paperclip by invoking the denial argument, noting that the influx of German scientists was an invaluable weapon in a Cold War that could turn hot at any moment. The Soviet Union began the postwar period with a comprehensive and aggressive program to simultaneously dismantle and exploit Germany's scientific infrastructure. Although the United States absconded with the vaunted rocket team, or at least most of the V-2's architects, the Soviet Union quickly produced comparable weapons and laid the foundation for its own MIC. The US intelligence community worked to secure German scientists in the Western occupation zones, recruit and lure scientists away from the Soviet zone, and acquire accurate reporting on Soviet utilization of foreign experts in defense fields. The United States relied on Germans for each of these missions and struggled to distinguish truth from obfuscation and self-promotion. The JIOA and the armed services enlisted Germans to provide meaningful anecdotal evidence concerning Soviet intentions and capabilities, all of which served to justify Paperclip and its successor programs.

CHAPTER FIVE

Paperclip Vindicated

German Scientists and the Maturation of the National Security State

> To pursue his lifelong dream, he has helped Adolf Hitler wage a vengeful new kind of war, has argued against bureaucracy in two languages and campaigned against official apathy and public disbelief through most of his adult years. . . . Von Braun's critics say he is more salesman than scientist; actually, he learned through the bitterest experience that his space dreams had to be sold ("I had to be a two-headed monster-scientist and public relations man")
>
> —"Reach for the Stars," *Time*, February 17, 1958

The decade between the end of Paperclip's acquisition phase in September 1947 and the launching of Sputnik in October 1957 seemingly vindicated the German scientists program. The contentious immigration issue and parade of embarrassing revelations related to Paperclip dissipated in the wake of increased Soviet aggressiveness in Europe, the testing of a Soviet bomb, and, most significantly, the surprise invasion of South Korea in June 1950. As the national security state mushroomed, exemplified by the return of limitless defense spending, the cadre of German scientists working across the spectrum of scientific endeavor assumed enormously influential positions in both private and public sectors. Now citizens of their adoptive country, the Paperclippers skillfully negotiated the confines of the national security state armed with security clearances, civil service contracts, and highly paid positions in the burgeoning defense industries. Von Braun's rocket team is certainly emblematic of this trend, but hundreds of other Paperclippers capitalized on their unprecedented access, courtesy of forgiving and generous clients. No longer "Prisoners of Peace" living in a state of glorified military custody, the

Paperclippers spent the first decade of the Cold War overseeing advanced research and development projects and cultivating the role of the scientist as advocate, a position many found familiar after performing similar functions in the Third Reich.

This chapter addresses the maturation of the national security state from the perspective of the Paperclippers, the intelligence and military officers invested in their success, and the civilian bureaucrats and business interests who increasingly networked with the German scientists. The first section concerns the long-standing civilian and commercial interest in Paperclip beginning with the National Interest directive allowing specialists to work in industry. The military eventually enlisted the Commerce Department's aid in placing scientists whose contracts with the armed services had expired. The JIOA encouraged companies to hire and sponsor scientists' immigration in part to circumvent the bureaucratic obstacles, but the JIOA faced new challenges with the creation of an independent Federal Republic of Germany (FRG). The second section of this chapter describes US actions in Germany during the Korean crisis. Fearing Soviet aggression in Europe, the JIOA initiated Project 63 to continue acquiring scientists for the MIC while denying them to a Soviet enemy poised to overrun the continent. The program violated German sovereignty, reignited the JIOA's feud with the State Department, and provoked another conflict with the CIA. In the third section, I revisit the plight of Paperclip's critics, specifically Samuel Klaus and the minority in the State Department who dared question Paperclip at its inception. Senator Joseph McCarthy and his allies savaged the State Department for their alleged crimes of sabotage and espionage, invoking Paperclip in their conspiratorial indictment of the civilian bureaucracy. Klaus ably defended himself, once again, but the idea that the State Department had undermined national security by obstructing Paperclip endured. Finally, I evaluate the success of Wernher von Braun's cohort in the United States. Handsome, articulate, and deemed invaluable by his patrons, von Braun crafted a powerful and effective message calibrated perfectly for an audience attuned to the national security imperative. From colorful spreads in national magazines to Walt Disney's studios, von Braun—the former SS officer responsible for building Hitler's "wonder weapons"—became the face of heroic science in Cold War America.

Paperclip and the Private Sector

Commercial interests influenced the exploitation of the defeated Reich from the moment the T-Forces hit the ground in late 1944. Some intrepid businessmen even accompanied the units as they swooped in on intriguing sites rumored to contain advanced technology and the brains behind them.[1] Vannevar Bush encouraged a high degree of civil-military cooperation by citing among the spoils of war "German technical information of an industrial nature." Bush envisioned using German data to maintain America's competitive advantage and provide employment for discharged veterans.[2] Bradley Dewey, a representative of the American Chemical Society, contacted secretary of state James Byrnes in summer 1945 to remind the government that "[r]esearch is always expensive" and that raiding Germany could save American companies untold time and resources. "Would that I were running a company with money enough to really wade in," Dewey wrote, "I would leave right now for Germany."[3] President Truman fielded requests from executives seeking access to the captured information and, increasingly, German personnel to fill open positions in their companies. Initially skeptical, Truman changed his views once he embraced the national security argument presented to him by Paperclip proponents. Yet, the rules governing commercial involvement were indistinct. The Meader Special Senate Committee investigated rumors of businesses profiting "directly or indirectly" from the military's occupation of Germany. The final report concluded that "there is some evidence of rather intimate business and social relationships between leading US businessmen and those in many German trusts. This, likewise, might have a bearing upon the vigor with which German industrialists are either prosecuted for war crimes or are subjected to the denazification procedure."[4] Embroiled in the bureaucratic battle with the State Department and a cynical public, the JIOA enlisted the aid of private industry and the Commerce Department to lobby on behalf of Paperclip and eventually provide a refuge for scientists the agency could not place in the armed services. The JIOA and some of the more astute Paperclippers recognized the symbiotic relationship between commercial and national security interests inherent in the German scientists program.

Henry Wallace's Commerce Department supported enthusiastically a broad exploitation program and pushed Truman to authorize Paperclip in November 1945. Wallace and his head of the OTS, John C. Green, ensured that Paperclip included a National Interest category for those scientists whose expertise could benefit civilian industry. These scientists were officially military employees who already had priority immigration status thanks to expedited security investigations, but they worked for an "industrial concern, educational institution or other civilian agency."[5] The OTS managed the myriad scientific consultants investigating German industry, but its stated mission concerned "putting the information in the hands of industry and the public."[6] As the Commerce representative to JIOA and a supporter of the denial strategy, Green supported bringing "outstanding scientists to this country on a permanent basis as an acquisition to our scientific talent."[7] Industrial investigators, Green discovered, exhausted the utility of interviewing Germans in custody abroad because the scientists performed best surrounded by similar talent and resources. Green cajoled his counterparts at the State Department's Economic Affairs section to help JIOA import "German scientists for permanent acquisition."[8] While the JIOA insisted on a "weapons first" policy regarding Paperclip acquisitions, the military invoked "national interest" instead of "national security" in the press release revealing the contours of the program to the public.[9] "It will be the government's policy that these Germans are exploited on behalf of the whole nation and not for or by single private interests," the text read. "In some cases arrangements will be made with industrial associations or societies for exploitation so that all those engaged in a particular industry may profit on an equal basis." The release concluded with the same argument Wallace used to convince Truman—the Germans constituted "intellectual reparations" whose value "will probably far exceed any cash reparations."[10] In 1947 Green designed a campaign for US industry entitled "Last Call for Germany," in which businesses were encouraged to request German specialists. Green asked executives to remember that "to the victor belong the spoils."[11]

Military officers responsible for Paperclip appreciated the benefits to civil industry, but the armed services regarded the relationship with the Commerce Department in strictly utilitarian terms, especially during the first year of Paperclip, when the mandate was limited and bureaucratic opposition fierce. For example, upon learning that a sponsoring agency decided to repatriate a scientist rather than renew the contract, the JIOA asked Commerce "to determine whether or not the scientist can be exploited" in lieu of returning to Germany.[12] The relationship changed after Paperclip "terminated" the acquisition phase on September 30, 1947. The civil exploitation program provided the best avenues for recruitment and allowed the JIOA to continue its mission with greater freedom and flexibility. John Green highlighted reports of the dearth of qualified scientists in fields like aeronautics in memos to his superiors and counterparts in the bureaucracy, noting that industry looked to Commerce as its "agent'" to intervene on its behalf and to expedite the Germans' paperwork with the State Department. Green suggested matching scientists' skills and abilities with openings in industry and proactively certifying their qualifications and security evaluations.[13] Ray Hicks, Green's colleague at the OTS, warned Green a week before Paperclip's termination that even if the program fulfilled the mandated one thousand scientists Truman authorized, it "would appear to be conservative in filling in the gaps which exist at the present time." Hicks anticipated more, not less, requests from industry after the program ended and urged Commerce to take the lead in publicizing scientists' availability and streamlining the "cumbersome" vetting process.[14]

Despite the OTS's lobbying, the National Interest program fell by the wayside because it originally expected to host fifty exceptional scientists working in civil industry, not hundreds. The private sector remained a useful outlet for the JIOA, however, even without Commerce's active cooperation. The OTS continued to aid JIOA, but Averell Harriman, Henry Wallace's successor in October 1946, balked at helping the military situate Paperclippers in civilian positions. In a memo to Secretary Byrnes in January 1947, Harriman wrote that he is "finding many new and intriguing activities within the Department of Commerce which I did not earlier suspect," namely the "joint War Department–Commerce Department endeavor for the collection and exploitation of German technology."[15] Harriman appreciated the value of microfilming tons of captured documents and extended his support to FIAT and military governor Lucius Clay, but Harriman understandably hesitated to sponsor an indeterminate number of Paperclippers, some of whom were of dubious value. Wallace and Harriman expected Nobel Prize winners like Werner Heisenberg and Gustav Hertz to grace the United States and fill high-profile academic positions, not scientists the military considered "overflow" or simply denial risks.[16] Validating Commerce's fears, AAF officers at Wright Field recommended the following categories of specialists for "immediate assignment to the Department of Commerce":

1. Those whose talents and abilities were not of long-term use to the AMC or its contractors.
2. Those who were trouble makers or whose presence affected the morale or working efficiency of the rest of the group.
3. A few whose character, loyalty, or general attitude made them undesirable regardless of their abilities.[17]

Given this admission, it is no wonder Harriman sought distance between Commerce and the armed services regarding Paperclip, but he was obligated to facilitate the program whether he liked it or not. John Green assured JIOA that Commerce "would do all in its power to assist the military in finding industrial employment for those German scientists that the military felt should be removed from Germany for reasons of military security." The denial strategy won the day once again. Harriman insisted, however, that Commerce restrict its role to assisting the military and abandon the notion of "co-sponsorship."[18] Between 1946 and 1948, the approximate dates of Harriman's tenure, Commerce sponsored zero Paperclip specialists. After 1948, the number reached fifty-two, the estimated number Wallace had intended with the National Interest program.[19]

Harriman's reluctance notwithstanding, numerous companies, universities, and congressmen attuned to the economic benefits of the scientist program contacted OTS with "wish lists" and other requests for assistance navigating the bureaucracy. In 1947 congressman Lyndon B. Johnson, who would shower former Paperclippers with resources and praise as president, inquired about a German refrigeration expert for a position at the University of Texas.[20] An executive at the Stewart-Warner Corporation wrote assistant secretary of war Howard Petersen that companies like his expected the government to "help cut the red tape" and "salvage any German brains out of this mess." The executive did not expect a blanket commitment, writing that he had "no sympathy with the chap who wants to get a couple of mechanics to fix his punch press or washing machine. But I do have sympathy with the university where they need a professor."[21] The Bulova watch company, on the other hand, regarded Wright Field as a private labor pool. Bulova complained to two US senators that the Germans were forced to work behind "barbed wire" as though they were working at a concentration camp. Secretary Petersen was alarmed to learn that "Mr. Bulova himself had taken an active part in getting three of these men brought to this country by the Army" and had used his influence to release them to the watch company.[22] The Sam Wood production company went so far as to direct the JIOA's collection and vetting efforts. Wood asked the army to secure seven scientists still on the continent, bypassing all pretense of security and immigration protocol. The army recommended that Wood "convince the Director of Intelligence . . . that the services of these men were needed by the Army" (even if they were not), while the JIOA volunteered to collect "data on German sound film and color processes" and urge Commerce to secure the equipment and personnel for Wood.[23] Henry Wallace's support for "intellectual reparations" from the defeated Reich helped convince Truman to authorize Paperclip and empower the JIOA in its struggles with the bureaucracy. More important in the long run, the private sector helped prolong Paperclip and facilitated the Germans' integration into the national security state.

The JIOA pushed for a more robust "civilian exploitation program" in March 1947 in part to evade State Department immigration procedures. JIOA chairman Bosquet Wev touted "limited military custody" for scientists loaned out to industry. "If we can encourage civilian agencies to accept and exploit these scientists," Wev wrote, "the US can receive the benefits of very important and valuable research which would otherwise be lost to the US and most probably be gained by foreign nations and competitors."[24] The JIOA used the OTS to find homes for scientists with expired Paperclip contracts in industry and at the growing number of universities dependent on government contracts, but promoting the scientist program was more difficult than expected. One memo discussing scientists' compensation expressed concern that industrial leaders, universities, and research labs were reluctant to pursue Paperclippers because they believed doing so alienated the armed services. Placing scientists outside the military "appears to require more 'selling' with more difficult and extended personal negotiations . . . than does their employment by the Armed Services on Paperclip projects."[25] Wev asked OTS to help publicize the "value to the national economy of the Paperclip Program" by chronicling the "broad effects of the specialists." Wev requested "[q]uotable statements from employers" and "analyses of . . . specific contributions by individual specialists" to justify acquiring scientists from what was now an independent German state.[26] Embittered by his dealings with Samuel Klaus and others in the State Department, Wev enlisted civilian partners in Paperclip to disarm critics who accused the JIOA of remilitarizing German science.

Some companies and universities rejected overtures by OTS because of the inevitable controversy involved in hiring German nationals with questionable pasts instead of qualified Americans. An executive from the Oklahoma Power and Propulsion Lab informed Wev that he did not appreciate the JIOA pushing obvious Nazis who "won't fit in here."[27] Carlton Murdock, the dean of the faculty at Cornell University, complained to the American Council on Education that "it is stated on one hand that [German scientists] have the status of prisoners of war and that they are returned to Germany when their usefulness is at an end," yet Cornell and others are urged to extend "permanent appointments" to specific scientists. Unsurprisingly, most of Murdock's faculty opposed hiring the Germans.[28] Although some universities understandably balked at the idea of hiring controversial Germans, even with the added incentive of military contracts, the JIOA counted on most civilian and private sector partners taking a more practical stance. Michael Straus, an assistant secretary at the Department of Interior, informed the JIOA that his interests were "confined to accession of technical and scientific knowledge from Germans, regardless of their democratic or Nazi point of view. . . . This department will not be interested in any social or governmental views of any of the Germans imported, regardless of whether they are Nazis or anti-Nazis." All that mattered, Straus continued, "is what we are to get out of the Germans. We do not expect the Germans will be getting anything out of this department."[29] Eventually, as Paperclip evolved, officials like Straus had to offer more to the Germans than an indifferent attitude toward their complicated pasts. As one senior military officer remarked in October 1948, "The days are gone when a carton of cigarettes was good enough for the preparation of a scientific report and for a CARE package one could have a copy of all the projected developments of a large factory."[30] Another official who managed Paperclippers at the Armament Lab at Wright Field noted that differences between scientists and technicians were greater in Germany than in the United States: "They [scientists] are having difficulty adjusting themselves to the American system without losing their dignity."[31] Once it became clear that the Paperclippers were permanent employees, not glorified POWs, both the military and the civilian sectors gladly paid them what they were worth, often more.

Project 63

Once Project Paperclip's acquisition phase ended in September 1947, the JIOA's mission transitioned to retaining the scientists as either civil service employees or private sector consultants safely removed from the imperiled continent. The JIOA's usefulness was in doubt in the late 1940s, but the combination of NSC-68, which recommended a diverse and systematic military buildup across the services, and the onset of the Korean conflict in June 1950 underscored the Paperclippers' potential contribution to the MIC.[32] The fast-moving events on the Korean peninsula raised the specter of a Soviet invasion of Europe and the capture of Germany's top scientists, but sending a second wave of recruiters was more complicated in 1950 than it had been in 1945. In June 1949, the State Department combined OMGUS, the ACC, and the US Group Control Council into one diplomatic office, High Commission for Germany (HICOG), working out of the new Bonn embassy. The JIOA, still operating as it had in occupied Germany, disregarded HICOG whenever it suited and willfully ignored the FRG's sovereignty over its own citizens and resources. In November 1949, the JIOA wrote the State Department's Office of German Affairs seeking clarification on a series of directives guiding HICOG responsibilities in Germany, ranging from the issuance of exit permits to surveilling "certain scientists and technicians."[33] Unconvinced HICOG could safeguard the scientists in Europe, the JIOA initiated a "new Paperclip" (Project 63) designed to transfer vulnerable scientists stateside. Project 63 damaged JIOA's relations with the military, the CIA, the FRG, and, once again, the State Department.

The JCS approved Project 63 in November 1950, "recognizing the desirability of establishing a program whereby certain outstanding German and Austrian scientists would be denied to a potential enemy." Often referred to as an "expanded Paperclip" or "accelerated Paperclip" in numerous memos, the JIOA was restricted to emigrating eighty-five Austrians and sixty-five additional Germans "on a voluntary basis" and offering candidates a six-month contract from the Department of Defense.[34] Fearful of a Soviet invasion of Europe in concert with North Korea's invasion of the South, the JCS acknowledged that Project 63 "is primarily designed to remove selected . . . scientists and specialists from areas subject to immediate conquest . . . and to utilize them in the national interest."[35] The program experienced delays, however, with the passage of the McCarran Internal Security Act in March 1950. The legislation clearly targeted Communists and other "subversives," but the provision denying citizenship to supporters of "totalitarian dictatorships" unintentionally included most Paperclippers and any future recruits from Germany and Austria.[36] The military swiftly inserted itself into the legislative process and carved out exceptions for the hundreds of ex-Nazis serving the MIC and intelligence community. The JIOA complied by scouring the Germans' personnel files for Communist affiliations, something it had steadfastly avoided when Nazi Party membership was the issue.[37] The FBI also dropped its objections to Nazi Party membership. Paperclipper Werner Ditzen laughingly told science journalist Morton Hunt, "The FBI didn't care about my being in the Nazi Party; I think they understand about that. What they wanted to know was whether perhaps I was a communist. It is funny, is it not? I, a communist!"[38] The JIOA reminded the civilian bureaucracy that returning scientists to Germany "would involve a serious security risk since many of them are thoroughly familiar with our research and development program."[39] Bosquet Wev argued that industry offered a safer alternative for scientists released from Paperclip projects, but a prosperous FRG complicated matters by offering the Paperclippers a meaningful future in their native land.[40]

HICOG designated Carl Nordstrom from the Scientific Research Division (SRD) to assist JIOA with Project 63, but there was a fundamental divide between the military and the State Department regarding the future of German science. The SRD shared the burden of vetting approximately 150 specialists, but Nordstrom was told not "to judge the acceptability of an individual or of an institution on any basis other than scientific competence."[41] Only five EUCOM military officers were assigned to Project 63, which exacerbated the communication problems between diplomatic and military officials. The JIOA regretted that "[c]onsiderable ill will has developed as a result which will make future cooperation between EUCOM and HICOG at this level quite difficult." Neither entity entirely understood the other's new mission in the FRG. "The military regard Project 63 as a military program," a HICOG memo notes, "and do not understand the activities of the operation managed by Mr. Nordstrom."[42] As a member of OMGUS, however, Nordstrom helped the JIOA acquire visas for what it called the "hot cases" and remained helpful to the program.[43] Nordstrom tried to reason with the JIOA regarding the best way to implement Project 63 and chastised them for disregarding Germans' "extremely sensitive" feelings and bypassing his office, which was in a better position to approach potential recruits.[44]

The JIOA understood that the playing field in Germany differed considerably since the agency had first begun deploying teams in 1945 and 1946, so it altered its tactics, specifically by involving successful Paperclippers in Project 63. In a paper entitled "Suggested Procedure for a New 'Paperclip' Project," the JIOA recommended building "an administrative organization made up of Germans. In this way from the first moment there will be confidence between the German who is looking for a job and a representative of US industry." Germans should be the face of the program and assimilate new recruits.[45] Officers working with Paperclippers at Wright Field recommended that the JIOA contact influential figures in German universities, including Werner Heisenberg, and "thoroughly indoctrinate them in all phases of the program. . . . The key personages exert sufficient influence over their associates to counteract any adverse criticism of the program which may originate from other sources."[46] Ernst Stuhlinger, chief of the Research Projects Office at the ABMA at Redstone Arsenal, visited Germany on a similar mission in 1955. As one officer noted in authorizing funds for the trip, "Past experience has proven that an American recruiter is more successful when accompanied by a former Paperclipper."[47] Hubertus Strughold, for example, helped recruit seven specialists in aviation medicine during a trip to Germany in July 1949, and Walter Dornberger accompanied "Pentagon brass" in a 1952 recruitment trip.[48] Using Germans to recruit other Germans worked reasonably well for the Soviet Union until the rumors of mistreatment and underemployment became too widespread to deny, but the idea never progressed within the JIOA beyond the occasional personal letter or visit. A 1955 memo lamenting lost opportunities relating to Project 63 acknowledged, regretfully, "We are dealing with humans that have minds of their own."[49]

Project 63 was a victim of the FRG's success and the JIOA's continued failure to cooperate productively with civilian counterparts. New and old critics of Paperclip condemned the JIOA for ignoring protocol and established procedure and blamed the program for undermining US efforts to develop a peaceful and autonomous German scientific community. The SRD worked closely with German universities and research labs, and several junior employees protested vehemently to Nordstrom, who had the unenviable task of implementing contradictory goals—Project 63 and German independence. Matthew Miller, one of HICOG's scientific liaisons in Germany, called the first Paperclip "a most serious mistake" and could hardly believe it was "still alive . . . especially when such strenuous efforts and money expenditures are being made by the American people to build Germany back to the point of self-sufficiency. I shudder of the consequences if [Project 63] is allowed to continue particularly on scientific, or otherwise, relations between Germany and the US."[50] Another SRD representative, John Edsall, thought keeping Project 63 acquisitions "permanently in America . . . may have a catastrophic effect on German science and could also do great damage to German-American relations." Absconding with "eminent scientific leaders" damages the morale of those who remain, Edsall wrote, citing the example of the Free University in Berlin to prove his point. Edsall had pumped resources into reestablishing Germany's prominent universities on a solid democratic footing, but Project 63 had the effect of building up "Berlin science and the FU [Free University] with one hand, while tearing them down with the other." Edsall recommended providing Germans fellowships to study or teach at American universities, especially after a decade of isolation, but Project 63 "threatens the normal development of German scientific life."[51] Edsall was not exaggerating the JIOA's vision for Project 63, which envisioned the wholesale removal of Germany's most valuable scientists. The JIOA informed EUCOM in September 1951 that "[s]ince Project 63 is based on the denial aspect of the program, a return clause should not be included if at all possible."[52]

Nordstrom's junior colleagues liaising with their German counterparts obviously had reason to protest Project 63, but more senior officials in the State Department and intelligence services feared that JIOA risked established relationships with Germany by aggressively recruiting its best scientists at a time when they were most needed. The State Department's Intelligence Office attacked the JIOA for its continued "free-wheeling" and "dangerously inept handling of the program." B. R. Shute, a special assistant in the office, considered it outrageous that JIOA could "propose a plan for assuring [*sic*] scientific intelligence rendition" with no jurisdiction. Shute concluded, "The Department should, in my opinion make clear to JIOA that Nordstrom's actions are controlled by the Department and not by that agency."[53] The CIA also resented the JIOA's careless recruiting efforts, noting that Project 63 "could conceivably impair [CIA] sources of intelligence should key scientists be withdrawn."[54] A 1952 CIA memo (with redacted names) related a meeting between several agencies regarding Project 63. The CIA representative noted that the JIOA's list of names included "people too old, too rich, too busy and too thoroughly disgruntled with past experience with Americans to justify a renewed approach." The CIA then "raised the question of propriety in connection with the use of ex-Germans as recruiters having in mind resurgent German nationalism and the possible attitude among present-day Germans that the Paperclippers are in effect traitors." The CIA was frustrated with the JIOA's blundering and inability to cooperate, noting that the military came around only after "[t]he storm we kicked up by our cables. . . . The slow awakening seems to be characteristic of the breed."[55] The State Department was not the only civilian agency exasperated with the JIOA.

Shortly after the creation of the FRG, the State Department assigned "scientific attachés" to establish long-term relationships with Germans in science, a move intended to emphasize Germany's independence from US oversight and the end of exploitation.[56] The attachés reported numerous awkward instances when Germans were approached by JIOA recruiters. One group of German physicists "couldn't understand this offer and, from a scientific point of view, it was of no help to them." Asking renowned scientists to abandon their academic posts for six months, possibly forever, for an unclear future in the United States was hardly an attractive offer, and thanks to German independence, it was an offer few chose to accept.[57] J. B. Koepli, the science adviser in the State Department, did not oppose "the merits or the objectives of [Project 63]," but "the manner in which it was carried out led to some very unfortunate repercussions within Germany and was a cause of considerable embarrassment."[58] The JIOA claimed the "morale of Project 63 contractor personnel and their dependents is exceedingly high," thanks to excellent housing and employment opportunities, but the weight of the evidence suggested otherwise.[59] Koepli contradicted the JIOA in a memo, reporting that "great dissatisfaction has been expressed by a number of the German scientists who have been brought to this country under contract. These individuals have communicated with friends and associates in Germany," which led to criticism of the United States and provided opportunities for those "who wish to damage our relations with West Germany."[60] State Department officers in Frankfurt predicted that Project 63 would provoke a "strong anti-American reaction" in the West German press and "furnish excellent material for Communist propaganda."[61] These fears were realized in 1951 when the Foundation for German Scientists published *Forschung Heisst Arbeit und Brot* (Research means work and bread), a book that condemned both Soviet and American exploitation operations for sabotaging Germany's future. The authors asked, "will the German scientists be able to live among us permanently if their own fatherland does not secure for them the foundations of their work?"[62] Project 63 never fulfilled its specified mission, but the original Paperclippers enjoyed tremendous earnings, prestige, and responsibilities in the national security state.

Samuel Klaus and the Red Scares

Project Paperclip weathered bureaucratic confusion, turf battles between competitive armed services, opponents in the State Department, and a broader public hostile to the program's controversial immigration clause. The JIOA considered the mandated acquisition phase spanning 1945 through 1947 a partial failure because it failed to recruit the one thousand scientists authorized by Truman, but the agency continued to exfiltrate talented specialists from Europe well into the 1950s. More importantly, the JIOA succeeded in retaining its initial recruits in the United States at a time when the national security infrastructure grew exponentially, allowing these once enemy aliens to occupy prestigious positions. The controversial scientist program was not just vindicated by the Cold War; it became a Cold War weapon with which to bludgeon anyone perceived "subversive" or "weak" on communism. Senator Joseph McCarthy's assault on the State Department encouraged his allies to rehash the Paperclip episode in such a way that Samuel Klaus assumed almost superhuman abilities to destroy America's national security single-handedly. Project Paperclip was suddenly reincarnated as a heroic chapter in the Cold War—

certainly not amoral, opportunistic, or even just a rational cost-benefit analysis, but an inspired and vital weapon against the Soviet peril. Paperclip's stunning change in perception matched the United States' speedy descent into Cold War paranoia. The McCarthyites, aided by FBI director Hoover, mounted a coordinated campaign against Samuel Klaus, a junior official. Klaus was targeted not simply because he supposedly hindered Paperclip's progress, but because he defended his colleagues from unwarranted FBI investigations and congressional "witch hunts" intended to weaken the State Department permanently.

When JIOA chairman Bosquet Wev testified before Congress in June 1947, he elicited political support for Project Paperclip by denigrating Samuel Klaus's obstructionism. Wev and JIOA director Thomas Ford supplied sympathetic politicians with a compelling portrait of Klaus and others in the State Department as veritable enemies of the state. Shortly after Wev's testimony, Senators Styles Bridges and Joseph Ball publicly denounced the State Department's handling of Paperclip, which helped cast doubt on department loyalties for years to come. Senator Bridges's July 18, 1947, appearance on *Meet the Press* concerned military preparedness and featured this remarkable exchange with journalist Laurence Spivak:

> MR. SPIVAK: Senator, just the mere expenditure of money doesn't mean that we're getting the kind of quality of equipment that we need. I understand, for example, that the Russians have taken away from Germany a great many important leading scientists, especially those who had something to do with jet propulsion. Are you familiar with that at all?
>
> SENATOR BRIDGES: I am, Sir, and I am very disturbed about it. That's one of the things that I'm terrifically concerned about. I might say that the Appropriations Committee, a subcommittee of the Appropriations Committee, has been holding some confidential hearings over the last, course of the last few weeks, looking into action by the State Department and State Department officials on various matters, and one of the things that vitally concerned us is the fact that America hasn't secured her share of the German scientists, while Russia has picked up a larger number and is utilizing them to the full extent.
>
> MR. SPIVAK: Do you think the State Department is responsible?
>
> SENATOR BRIDGES: I don't know who the roadblock is. We're looking into that to try to find out where to put the blame and try to remove the roadblock, but it certainly occurred somewhere in the State Department.[63]

The notion that the United States lost German scientists the same way it "lost" China to communism morphed into conventional wisdom as Cold

War tensions hardened. Bridges floated the theory in 1947, but by 1950, it somehow became an article of faith that the State Department, either through incompetence or deliberate sabotage, had endangered national security by requiring a vetting process for German scientists.

Fortunately for Klaus (and historians seeking insight into this tumultuous period), the besieged lawyer wrote extensive memos recounting every meeting and phone call related to his Paperclip ordeal. Reading both Klaus's correspondence and the FBI files concerning the State Department makes clear that Paperclip offered the department's enemies an opportunity to malign and strip Truman's State Department of a meaningful role in the national security state. Klaus consistently noted his JIOA assignment was an additional duty unrelated to his position in the State Department's Security Office, yet he was targeted for his work in both positions. The FBI, and Hoover personally, had long-standing grievances against Klaus predating World War II. Klaus, whose unctuous personality frequently alienated his colleagues, exacerbated existing contempt for the State Department among the FBI, Congress, and the military.

The weeks preceding Bridges' *Meet the Press* interview were busy ones for Klaus and his detractors. Klaus's memos indicate that personal meetings with Senators Bridges and Ball and several staffers went surprisingly well, yet the senators attacked Klaus as soon as they left the room. Klaus soon discerned the root causes of his public defamation—embittered State Department officials he had fired or investigated and the newly converted Paperclip enthusiast J. Edgar Hoover. Hoover had first learned of Klaus in 1940 from the US attorney in Chicago when Klaus had worked for the Treasury Department as an investigator. Hoover recalled the attorney informing him that Klaus "is a man of extreme enthusiasm and energy, and he [the US Attorney] wished to caution me that he might try to run the whole FBI if he was permitted to do so." Although said in jest, the exchange resurfaced in Hoover's memory years later in a different light. Hoover had written his subordinates in 1940 that he desired "a cordial and close relationship with [Klaus]. It should be kept in mind that the Secretary [Hans Morgenthau] indicated he may be a rather difficult person to work with."[64] Klaus worked effectively with the FBI on the Safehaven program, but the relationship deteriorated in summer 1946 once Klaus joined the State Department and oversaw personnel investigations for Joseph Panuch, the deputy assistant secretary of state for administration. This sensitive assignment coincided with the JIOA appointment. Klaus essentially overhauled the State Department's security procedures in less than a month, dismissed experienced investigators, and refused to cooperate with the FBI unless the agency revealed all its sources relating to personnel investigations.[65] George C. Marshall dismantled Klaus's office, the Advisory Committee on Personnel Security, in June 1947 and replaced it with the controversial "loyalty program." This occurred soon after Klaus's removal from the JIOA. According to the State Department's official history of the Bureau of Diplomatic Security, Klaus did real damage while overseeing security matters, leaving the department divided and vulnerable to the likes of McCarthy.[66]

The politics of personnel investigations and espionage in the State Department are beyond the scope of this book, but Klaus's exchanges with the FBI and Congress are linked to Paperclip's vindication in the age of McCarthy. Hoover coveted powers allocated to the State Department, such as issuing visas to foreign visitors, and predictably conspired against this rival.[67] Hoover dropped a bombshell on Truman in May 1946, informing the president of "an enormous Soviet espionage ring in Washington" comprising dozens of senior bureaucrats, including two State Department officials connected to Paperclip—undersecretary of state Dean Acheson, Klaus's mentor, and John J. McCloy, the US high commissioner in Germany.[68] As part of the investigation into the supposed espionage ring, Hoover personally ordered "a most discreet inquiry into the present activities and nationalistic sympathies of Samuel Klaus," just weeks after Klaus issued his negative report on the FBI's influence on State Department investigations. Hoover demanded information on Klaus's associates, his family, and why he "traveled extensively in Europe and on June 26, 1933, received a passport valid for travel to Soviet Russia." Hoover concluded that "[g]reat care should be exercised to avoid revealing the Bureau's interest in Klaus."[69] Hoover received several reports highlighting Klaus's membership in the Anti-Defamation League and his sister Ida's attendance at a dance sponsored by the Washington, DC, Women's Committee to Aid Children of Spanish Democracy. Klaus demanding the FBI share information enraged Hoover, who wrote in the margins of one report that "Klaus might very well want to know *all* in view of his own subversive connections. He should get nothing."[70] Klaus had expected the scrutiny and learned from his contacts that Hoover had accused him of revealing secrets to the press, but apparently Klaus's family connection to labor leader David Dubinsky was enough for Hoover to "lay off." Hoover finally relented, stating, "everything was ok so far as I was concerned." Klaus worried about the illegal disclosure charge, but his contact assured him he was known to Hoover as "Top Secret Klaus," and the most Hoover could say about him was that he was overly critical of the FBI.[71]

Klaus evaded personal destruction, but his Paperclip protests remained a point of contention. Klaus's memos reveal that his congressional critics had unfettered access to FBI, JIOA, and State Department files. John Peurifoy, Klaus's superior, told Klaus about the Paperclip investigations in Congress and Senators Ball and Bridges' claims that Klaus "had worked in the Russian cause because," as Klaus described it, "I had kept security cases from being decided and because I had gone against the instructions of the department in preventing German scientists from being snatched from the Russians."[72] Klaus and Peurifoy linked the Paperclip criticism to the ongoing personnel security investigation controversy. Klaus met with Senator Ball the next day and related his side of the story. Ball, according to Klaus's memo, "said that he had never asked for my dismissal but he had simply pointed out that I appeared to be the person . . . who was responsible for the Department's taking a pro-Russian stand." Ball insisted it was not "his business to interfere" in departmental affairs, but "a former disgruntled employee" told Ball that Klaus was "a communist and was protecting communists in the department."[73] Klaus followed up with several staffers and Peurifoy and determined that Bosquet Wev's testimony on June 27 had supplied Bridges and Ball false "evidence that I had personally opposed the program which had the support of the President; and that the order of the President permitting immigration superseded the immigration laws."[74] Klaus met privately with Bridges just four days after the *Meet the Press* interview and learned that Bridges had also heard from anonymous sources that "I was working in the communist cause. We agreed that if this was true I should not only be removed from my job, but I should be indicted and prosecuted. He said that the two subjects that had been discussed had been my work in the personal security field and the German scientists program."[75] Both senators were willing to drop the Communist issue, but Senator Ball in particular "was not satisfied about the handling of the German Scientists Program."[76] Klaus may have convinced himself this storm had passed, but the accusations returned in force in 1948 and again in 1950.

Shortly before his first episode with Congress in spring 1947, Klaus wrote a memo indicating that George C. Marshall had personally vouched for State Department employees "to the members of the House and Senate committees covering Foreign Affairs, State Department appropriations, and Un-American Activities." Klaus assumed correctly that nothing would convince

Congress of "the loyalty and sincerity of the departmental officers."[77] In spring 1948, Republican congressman Fred Busbey used his position on a subcommittee of the Committee on Expenditures to investigate Truman's bureaucracy for "disloyalty."[78] Busbey gathered all the information available on Klaus from State Department files. He accused Klaus's Security Office of harboring "a group of pro-Communist fellow travelers and muddleheads," and submitted a lengthy tirade to the *Congressional Record* entitled "What's wrong with the State Department?"[79] The statement echoed Wev's testimony, but Busbey accorded Klaus more power than even the JIOA implied:

> That from the first instance Klaus has used his authority as the State Department representative on the JIOA to delay, obstruct, and confuse the program. He has even gone so far as to state this [Paperclip] may be the policy of the Secretary of State and the President, but it is not the policy of Samuel Klaus. . . . The influence which Klaus wields in the State Department is unbelievable. Wherever the War and Navy Departments touched in an attempt to unravel the confusion is found an official who could not act or say a word until he contacted Klaus.

Busbey also accused Klaus of "obsessing" over the DP program, and though Klaus allowed fifteen hundred poor refugees into the country, "he could not arrange for even one scientist to come to the US to be placed in a program that was of utmost interest to the US."[80] Busbey accused multiple officials of being Communists, including the congressmen on his committee, inserting the statement shortly before resigning in protest.[81] Busbey blamed committee chairman Edgar Chenoweth of closing the investigation "before all the facts could be brought to light."[82]

Klaus immediately mined his extensive memos and refuted Busbey's statement point by point, noting, "The false, and fantastic, allegations and innuendos are so many that only the chief ones can be noted." Klaus's memo recounted every moment of his Paperclip assignment, complete with references to immigration laws, executive orders, JIOA minutes, and conversations with military officers who verified Klaus's accounts. Klaus reminded anyone who cared that he worked to make Paperclip a success by continually pressing for an expansive military custody program. He reiterated that he never denied visa applications because, he wrote angrily, "I continually requested of JIOA THAT INDIVIDUAL CASES BE SUBMITTED FOR ACTION."[83] Klaus was disappointed to learn that Chenoweth's subcommittee "considering loyalty investigations" had concluded before he could answer Busbey's allegations. Klaus wrote an abridged version of his original memo to Chenoweth, reporting that Busbey never "made any attempt to elicit the facts from me" regarding Paperclip, let alone the security investigations. If called to testify, "I would have demonstrated that not only are the derogatory allegations of Congressman Busbey untrue and in the main easily refutable but such participation as I had in that program [Paperclip] had the purpose and effect of preventing the unobstructed recruitment by unfriendly foreign powers of scientists and technicians" and "irreparable damage to the US." Klaus informed Chenoweth that Busbey's charges likely derived from Klaus's enemies at the JIOA. Klaus welcomed the "undisclosed individuals'" public testimony so that they "may be questioned and cross-questioned, their reliability ascertained and their true motives laid bare."[84] Klaus's skillful and thorough repudiation ended this second wave of attacks, but the third iteration placed Paperclip amid the dangerous summer of 1950.

On February 9, 1950, Senator Joseph McCarthy delivered a now infamous five-hour speech at Wheeling, West Virginia, exposing an alleged Communist conspiracy deeply rooted in the US government. Echoing Hoover's allegations made privately to Truman years earlier, McCarthy singled out the State Department: "There the bright young men who are born with silver spoons in their mouths are the ones who have been the worst. . . . In my opinion the State Department, which is one of the most important government departments, is thoroughly infested with Communists." Secretary of state Dean Acheson, that "pompous diplomat in striped pants, with a phony British accent," was McCarthy's primary target.[85] Acheson wrote in his memoir that Senator Styles Bridges leveled charges against him, John Peurifoy, and the Security and Loyalty Board. Acheson remembered the years 1950 through 1953 as a "slaughter" and a protracted "night of the long knives."[86] Given Samuel Klaus's connections with most of the people involved, it was only a matter of time before the beleaguered lawyer joined the fray. The Subcommittee on the Investigation of Loyalty of State Department Employees, the so-called Tydings Committee, pursued McCarthy's allegations and consumed the lives of its targets, Klaus included. A March 17, 1950, State Department personnel memo noted Klaus "was fully engaged in work connected with the McCarthy hearings and will probably continue to be for an indefinite period, perhaps running to several months."[87] Derisively labeled a "protégé of Acheson" in 1947, Klaus had years to anticipate which knives were thrown in his direction.[88]

The Tydings Committee focused primarily on the State Department's broken security infrastructure, but the surprise invasion of South Korea in

June 1950 resurrected the Paperclip debate in the context of military effectiveness. On July 18, 1950, Senator Bridges and Senator Homer Ferguson, both Republicans, rehashed a litany of charges against the State Department, beginning with the China White Paper, "a monstrous excuse for losing the key to the Orient," before settling on Samuel Klaus's disastrous stints on both the JIOA and the Security Office. Alarmed at North Korea's victories, Bridges tied the military situation on the peninsula to Klaus's Paperclip obstructionism, although he did not use his name on record: "I have no doubt this is the so-called German scientist program coming home to roost," Bridges claimed, citing old testimony that "the program was less than one-fourth effective. This man [Klaus] boasted that the scientist program might be approved by the President of the United States and the Cabinet, but it was not approved by him. He stopped it. The German brains went to Russia. Now the Russian tanks have impenetrable armor. And that man is still employed in the Department." Bridges capitalized on the crisis atmosphere, noting that the "Cold War is over. This may be the opening skirmish of World War III," and that "one man" in the State Department had endangered national security. "The State Department needs a real house cleaning," Bridges concluded, invoking the worst McCarthyite rhetoric. "This is not a job of sweeping the dust under the rug with a whisk broom, or airing out the house by opening the front door and sweeping the front steps. This job calls for yellow soap, a scrubbing brush, and plenty of elbow grease from the basement to the attic. It should be finished off with a first-class cyanide fumigating job."[89] A few days later Ferguson read a letter from the 1947 hearings that had first placed Klaus at the heart of the alleged conspiracy to derail Paperclip.[90]

Klaus confidently and calmly responded to the barrage of attacks, scrupulously reconstructing the chain of events in such a way that even his accusers were forced to reconsider their accusations against him. Undersecretary of the navy John Sullivan vouched for Klaus and arranged for Bridges to meet with Klaus to hear his side of the story on July 21. Just as he had with Senator Ball, Klaus convinced Bridges that he was not the root of the problem and that the accusations derived from "a cabal of the discredited security men" from the State Department "and the two discredited military men [Bosquet Wev and Thomas Ford]."[91] After his meeting with the senator, Klaus wrote that Bridges "would not refer to me on this subject" again and would report back to his committee that "there were documents in my possession that would show exactly what had happened, and that it was not necessary simply to raise the questions of my own credibility."[92] Moreover, Bridges agreed with Klaus that the scientists should have been placed under military custody instead of relying on expedited visas.[93] Klaus learned a week later that Bridges "received the [July 18] speech which he delivered from the RNC [Republican National Committee] and had, in fact, no knowledge of what the speech contained until he read it on the floor." Whether true or not, it certainly explains why Bridges repeated old accusations even after meeting with Klaus. "The RNC is busy defending McCarthy," Klaus wrote, "and [RNC official] Vic Johnson is in charge of writing these speeches."[94] Klaus was forced to revisit Paperclip one last time in June 1951 in a conversation with an RNC investigator and an aide to Senator Bourke Hickenlooper. Once again, Klaus disarmed critics and deftly turned the tables on his shadow accusers: "I said I could not say who pushed these two men into peddling these stories, but that it was an old tactic of the Communists to try to discredit someone who is dangerous to them."[95] The issue of Paperclip was simply too valuable for the McCarthyites to abandon. Klaus may have escaped personal ruin, although it is difficult to determine how his career was affected, but the State Department never recovered from McCarthy and his allies' ceaseless attacks.

It is important to note how often Paperclip's defenders resorted to antisemitism, both implicit and explicit, when confronting Klaus and other critics. Several JIOA officers accused Klaus of sabotaging Paperclippers' immigration because he was Jewish, a sentiment several Paperclippers shared openly decades later.[96] Klaus wrote in one of his responses to congressional testimony that "never in all my career had I been subjected to such personal mistreatment as by the military group. . . . I referred to the reports I had of personal bigotry of these men, particularly Wev."[97] Seymour Rubin, Klaus's colleague during the Safehaven program and later at the State Department, addressed the department's complicated history with antisemitism in a 1997 interview. Rubin believed he overcame institutional prejudice in the 1940s because he "was a respectable Jew who could do things," but being Jewish also seemed to invalidate his criticism of programs like Paperclip.[98] When Rubin suggested that inviting hundreds of inadequately vetted Paperclippers into the country might inflame antisemitism in the United States, Claire Wilcox, a State Department economist, countered that refusing the specialists' entry based on nationality made Rubin "involved in the same sort of nonsense that the Nazis preached."[99] Rubin and Wilcox eventually repaired their relationship, but Rubin believed his opinions were overshadowed by his Jewishness in the minds of his peers.[100]

American Jews were inexplicably blamed for controversy surrounding the more notorious Paperclippers. Shortly after the German epidemiologist Walter Schreiber joined the staff at the Air Force School of Medicine in 1951, several press outlets reported his links to medical experimentation during the Third Reich, prompting the air force, CIA, and JIOA to facilitate his emigration to Argentina.[101] Allen Dulles personally handled the matter, agreeing that the United States could benefit from Schreiber's knowledge if he served a friendly regime like Argentina's.[102] Unfortunately for the air force, a group of American doctors released damning military documents and protested Schreiber's continued employment all the way to the White House. One embarrassing memo quoted an air force official who declared Schreiber "too hot for me" and blamed "an organized medical movement against [Schreiber], emanating from Boston, by medical men of Jewish ancestry, I would suspect."[103] Thirty-five years later, Arthur Rudolph's defenders also implied that his critics were bitter Jews who harbored anti-German sentiment. The truth was that the more the space program captivated the public, the more forgiving, or indifferent, it was to the backgrounds of the Germans who built it.

The Indispensable Wernher von Braun

The year 1950 was fateful for the Cold War. From Truman's approval of the NSC-68 recommendations to McCarthy's witch hunts and the shocking Communist invasion of South Korea, the United States experienced a combination of increased dread and resolve. These events accelerated the mobilization of science, both civilian and military, and elevated the status of the most prominent Paperclipper—Wernher von Braun.[104] Von Braun's rocket team comprised approximately 120 scientists, engineers, and technicians whose principal duties involved rebuilding and testing the original V-2 missiles in Texas and New Mexico. By 1950 von Braun's guided missile project assumed greater significance, especially in light of a Soviet atomic bomb. The intercontinental ballistic missile (ICBM) was championed by multiple services, Congress, and influential allies in the press. General Donald L. Putt, who had managed Paperclip projects at Wright Field as a colonel, promoted ICBMs from his new post commanding the Air Research and Development Command.[105] Von Braun felt vindicated but could not resist complaining about his first years in the United States: "Frankly, we were disappointed with what we found in this country in the first year or so. At Peenemünde we'd been coddled. Here you were counting pennies. Your armed forces were being demobilized and everybody wanted military expenditures curtailed."[106] In November 1950, the army consolidated the rocket program under the Ordnance Guided Missile Center and transferred von Braun's team to an expansive new facility at Redstone Arsenal in Huntsville, Alabama.[107] From his position as a celebrated scientist responsible for a vital defense program, von Braun skillfully navigated the incipient MIC, a skill he and Walter Dornberger had developed in the Third Reich. Von Braun's charisma and foresight enabled his team to transcend whatever stigma remained from the Paperclip controversies and to prosper personally and professionally.

Before von Braun could play an integral role in defense policy and preach the gospel of space exploration, he had to consolidate his position with fellow Paperclippers and military employers. As I discuss in chapter 1, the rocket team included factions and strident personalities, which challenged the von Braun–Dornberger leadership clique. As the first group of Paperclippers to enter the United States, the rocket team was vetted, disciplined, and managed by the US Army. Some team members discerned their importance better than others and advocated forcefully for improved conditions, compensation, and a swift resolution to the immigration question. Ernst Steinhoff, an original member of the Peenemünde group and a trusted von Braun associate, drafted a letter to Colonel Holger Toftoy delineating multiple grievances shortly before the team's transfer to Redstone Arsenal. Steinhoff likely conferred with army officers before writing the letter, and he knew that Toftoy shared his opinions about the immigration process, but it is telling that the Paperclippers' complaints were handled very seriously by military officers in charge of the program. Steinhoff, understandably confused by the original Paperclippers' exact legal status, dissected contradictions in the law while emphasizing the team's value: "I think we have discharged our duties and responsibilities to the armed services and the US government with no mental reservation and with utmost loyalty," Steinhoff wrote. Furthermore, the team paid taxes and "handled the education of our children in a way which facilitated their personal adjustment to the American way of life and which gives them all the basic prerequisites to become good and desirable American citizens." Steinhoff asked why team members received the highest security classification and were entrusted with important projects, yet they were denied citizenship on technicalities. Writing just weeks after the invasion of South Korea, Steinhoff concluded: "We are generally afraid that in case of a war, which seems imminent, right now, laws may be passed which will affect us in a very negative way as aliens. We ourselves feel that we are participating and

Officials of the Army Ballistic Missile Agency, at Redstone Arsenal, Huntsville, Alabama, February 27, 1956. *Counterclockwise from left,* Holger Toftoy (*standing*), who recommended the transfer of the rocket team to the United States in 1945; Ernst Stuhlinger (*seated*), an original member of the rocket team and later a NASA scientist; Hermann Oberth (*forefront*), a pioneer in rocketry and an early influence on von Braun; Wernher von Braun (*seated*); and Robert Lusser (*standing*), German engineer and aircraft designer who joined the rocket team in 1953. *Source:* NASA on the Commons, courtesy US Army.

contributing very much in the American effort for the security of this country and the Western culture."[108] Toftoy's superior, Major General E. L. Ford, emphatically agreed and circulated the letter broadly, which further suggests that Steinhoff coordinated the letter with sympathetic army officers.[109]

Von Braun had selected the rocket team from among hundreds of Peenemünde and Mittelwerk associates, with minimal interference from US authorities. Consequently, some members did not have the requisite knowledge for advanced technical work, but they demonstrated unfailing loyalty to von Braun personally. Several intelligence officers, beginning with the perceptive Walter Jessel, discerned von Braun's agenda and penchant for rewarding friends and denigrating competitors. Paperclipper Walter Häussermann followed von Braun to NASA but acknowledged that his benefactor fostered an inner circle and kept others at a distance.[110] Von Braun could also be vindictive. According to CIC reporting from Germany, the wartime rivalry between the German army's missile program and the German air force's investment in a remote-control antiaircraft rocket bled into the postwar era. Von Braun, Steinhoff, and Herbert Axster "have done what they could to misrepresent and discredit the work of Dr. [Günter Theodor] Netzer's research group." Von Braun likely feared that Netzer would draw resources away from the rocket team in the United States as well. The CIC report, possibly citing Jessel's analysis from June 1945, noted that "Dr. von Braun has two men in his group who are not scientists or technicians. These men are Dr. Axster . . . and Director Steinhoff. . . . It appears that Dr. von Braun may attempt to portray the backgrounds of his two associates . . . as eminently scientific. It appears reasonable that his information may be utilized in the consideration of the exploitation of these men in the services of the US."[111] Indeed, von Braun convinced the US Army that Steinhoff and Axster's backgrounds in business and patent law were integral to the team's success in Germany and in the United States.

Several team members resented that nonscientists managed their projects and exerted undue influence over the team's new life in the United States. Gerhard Reisig, a talented engineer, remembered Ernst Steinhoff as "incompetent" and "a big mouth" who stole credit from others. Reisig admired von Braun but thought "he was too loyal to his people. He should have thrown Steinhoff out, because he did so much damage to the organization."[112] Major James Hamill, the officer charged with handling Paperclip security investigations, tried to neutralize criticism of Axster's continued presence in the United States, even after the revelations about his wife's sordid Nazi past. Hamill reminded the army that Axster "is an unusually able and competent administrator" whose "enthusiastic compliance with the directives at this office has at times brought him disfavor from his associates." Hamill recognized that Axster was not a scientist but noted that the original idea behind Paperclip "was to bring an integrated and well balanced team to the USA. . . . Axster was one of those who was selected on this basis."[113] Of course, Hamill neglected to report that von Braun, not the US Army, was the one who insisted that the patent lawyer was crucial to his team's success.

At some point during the rocket team's stint at Fort Bliss, the relationship between von Braun and Axster deteriorated, probably because of the embarrassing press coverage and Axster's domineering personality. According to an October 1949 FBI report for which von Braun was interviewed, "Axster has become uncooperative and he [von Braun] would recommend Axster for a visa only if Axster could be separated from the project or from the influence of his wife." Von Braun blamed Ilse Axster for being "gullible to radical political ideas" like National Socialism and causing "the family to earn a reputation which is not considered desirable in the US." Von Braun reported that the Axsters "asked people in socially and in a very cunning manner [to] undermine the authority of the group leaders" by exploiting personal weaknesses or illness to make the families of team members "feel indebted to [Ilse]." Several colleagues told von Braun "that they had the feeling of a rabbit in front of a rattlesnake attempting to hypnotize them." Von Braun informed the FBI that despite Herbert Axster's abilities, his wife's influence rendered him "a definite hazard to the entire group, which must operate as a team." Von Braun understood that the JIOA prioritized denial and fought against returning any Paperclipper to Germany, but he clearly preferred that to granting Axster a visa. He assured the FBI investigator that Axster's "technical knowledge of rocketry was negligible."[114] The JIOA accepted the findings of the FBI (and von Braun) that Axster's "desire to supplant von Braun as leader of the scientists" combined with his ignorance of the technology indicated that Axster "is not indispensable to the Project and that he could readily be replaced by someone whose influence would be beneficial rather than detrimental to so important an operation as that being conducted at Ft. Bliss."[115] Von Braun wielded the same influence he had used to secure Axster a coveted Paperclip contract to depict him as a toxic and extraneous rival. Whether justified or not, the episode demonstrates von Braun's clout and autonomy over a priority Paperclip project.

Contrary to the revised JIOA files pertaining to members of the rocket team, the FBI conducted thorough investigations containing frank assessments of scientists' characters, personalities, and willingness to adopt to an "American way of life." Specialist Hans Brede apparently did not embrace his opportunity in the United States. According to FBI informants, Brede "continually makes disparaging remarks about USA. He calls the late President 'Rosenfeld.' He asserts that once he gets the visa and subsequent first papers, things will be different—we won't be able to push him around anymore." Brede mocked the educational system, suggesting teachers "are purposely underpaid because America wants to keep its people stupid—so that they will be obedient."[116] The military chose not to renew Brede's contract and prevented him from interacting with potential recruits in Germany to ensure he could not poison the well.[117] Brede never denied his statements, but the FBI reports prove that some of the more well-adjusted Paperclippers did endure false accusations about their activities as legal residents. One concerned Huntsville citizen claimed that two Germans working at Redstone Arsenal gave each other the Hitler salute and openly mouthed "Heil Hitler!" during a "busy part of the day."[118] The FBI quickly dismissed the information but invested considerable resources resolving similar baseless charges.

Most interactions between the "Huntsville Germans" and the rest of the city were overwhelmingly positive. The rocket team's transfer to the southern city marked a new phase in which the Paperclippers lived and worked as normal citizens.[119] Anti-German outbursts were apparently rare, but some citizens and coworkers resented the Paperclippers' perceived privileged status. In one FBI report, the investigator observed that military personnel particularly "resented taking orders from a German."[120] A navy veteran named Myron Huston informed the FBI that the Paperclippers at Redstone Arsenal were "freezing out the native-born Americans who are employed there," noting that the local Veterans of Foreign Wars post was "considering taking up this matter through its Washington representative with Congress." Huston stopped short of accusing the Germans of "subversive activity" but "just didn't feel right about them having access to top secret information and being given preferences to native-born American citizens."[121] Paperclip spouses often felt more uncomfortable than the specialists when interacting with Americans. Erich Traub's wife was, reportedly, "never happy in this country and . . . resented the feeling held by some in the US against the German people."[122] Most Germans, however, were received warmly in their new southern home.

J. Edgar Hoover may have embraced Paperclip as integral to the new national security environment, but his agents in the field performed their duties dispassionately. The FBI wrote candid assessments of Paperclippers, noting that even the elite rocket team contained a certain number of, to use Samuel Klaus's language, "drones" with a range of personalities. The FBI seemed to prefer the "typical scientist" demeanor. Dieter Grau was "considered very meek," "mild-mannered," and purely interested in work.[123] Erich Traub was "a typical German, 'cock sure and arrogant,'" but also focused on "all things scientific." Other informants, however, thought Traub was a devout Nazi incapable of becoming a loyal American, not an uncommon sentiment emanating from some of the Paperclippers' American coworkers.[124] Arthur Rudolph, on the other hand, was described as a "natural born Republican" who "does not adhere to any ideologies which are contrary to the ideals of American democracy."[125] Thanks to the JIOA and von Braun's coaching over the years, most Paperclippers were savvy interview subjects by the time the FBI got a hold of them. For example, virtually every Paperclipper expressed gratitude for living in a democratic country and serving Western civilization. Von Braun was usually interviewed as a trusted source about the team, which only guaranteed his influence on who stayed and who faced exile, either to some other industry or back to Germany. Magnus von Braun, whose first security investigations included the infamous gold bar theft and blatant pro-Nazi utterances, had apparently become "very 'Americanized'" in Huntsville.[126] One informant believed Magnus used to be "cocky and somewhat of a playboy" until his parents came to the United States. The father "is the dominant figure in the household and [Magnus] has been somewhat more subdued since his parents' arrival."[127] Wernher von Braun's unchecked influence over the composition and treatment of his rocket team is the one constant in both the FBI and the military files. The more von Braun produced results, or at least persuaded officers results were forthcoming, the less interference he could expect.

The rocket team was accustomed to working in an environment marked by budget shortages and fierce turf wars. The rivalry between the army and the newly independent air force over the future of the missile program resembled some of the Third Reich's internecine conflicts. Von Braun was the army's secret weapon, and he worked to secure a sizable budget for Redstone Arsenal, although other Paperclippers joined the air force's program, based in California.[128] In August 1950 Truman appointed Kaufmann Keller, the president of Chrysler Corporation, as "missile czar," to study the military value and to estimate costs. Keller's findings were mixed, but a more favorable report by Aristid Grosse, a chemical engineer who advised the Defense Department, relied on "many hours of conference with Dr. Wernher von Braun."[129] The

CIA warned that the "acute shortage" of guided missile experts in the United States impeded efforts to keep pace with the Soviet Union. "A long time and a great deal of money are required to train men having the necessary foundation for this work," the report noted. Consequently, the United States should recruit from the "large reservoir" of specialists still in Germany. The JIOA used this report to justify Project 63.[130] Unfortunately for the army, von Braun's reputation and success encouraged the air force to accelerate its own missile program. The army, like Peenemünde, built missiles "under one roof," while the air force preferred, in the words of ICBM architect General Bernard Schriever, "going to industry and having industry develop and produce for us." Companies like Boeing, North American, General Dynamics, and United Aircraft manufactured the air force airframes.[131] In the battle between a private sector approach and the "old in-house arsenal system" von Braun championed, four historians of the MIC conclude, "the Air Force won. By eschewing any in-house capacity, the Air Force commanded a big, well-heeled constituency of scientists, organized labor, and industry. Its impeccable free-enterprise viewpoint removed most of the stigma from the surge in public spending that the arms revolution entailed." Although von Braun's

Launch of the Bumper 8, July 24, 1950. Bumper 8 was a General Electric Company program for testing and research, using a WAC Corporal rocket on a V-2 missile base. The Bumper program, created by Holger Toftoy in July 1946, represents the successful integration of World War II-era German military technology with American engineering. *Source:* NASA/US Army

model competed well, especially at NASA, the inclusion of outside contractors in large defense projects outperformed the "in-house" paradigm. Eventually, NASA followed suit.[132]

From the Paperclippers' interrogations with Walter Jessel in 1945 to interviews conducted at the end of their lives, they exuded confidence, some might say arrogance, in both their technical prowess and their ability to translate blueprints into a finished product. Walter Dornberger wrote a prescient memo in 1948 resembling the tone and some of the conclusions reached in NSC-68 a few years later. The former Wehrmacht general and V-2 project manager urged the United States to invest heavily in military research and development: "Such a program must be set up even if its organization appears to violate American economic ideals and American traditions in arms development."[133] At Redstone Arsenal, the team reconstituted Peenemünde, minus the concentration camp labor, and limited contracting to subsystems and mass production.[134] Eberhard Rees, who ended his career as director of NASA's Marshall Space Flight Center, touted the Peenemünde model of "big shops" and bragged that "sometimes people from Washington didn't like that too much, but they couldn't do a thing about it."[135] Dieter Huzel worked at Siemens before joining Peenemünde during the war and witnessed the origins of the "rubbing together . . . of these two cultures."[136] Gerhard Reisig stated in 1989 that "when we came over to this country, we more or less continued Peenemünde." Reisig respected the US Army because "[t]hey didn't say they know everything better, because they didn't know much about rocketry, and they were very willing to learn." Contrarily, the civilians at NASA "were arrogant" and had to be put in their place by von Braun.[137] Reisig defended Peenemünde's "under one roof" approach and condemned "the philosophy of industry," which he believed benefited contractors more than it benefited the army. Reisig shrugged off criticism that the team "over-engineered" and claimed they delivered top products within a budget. "I don't want to brag about it, but imagine a vehicle like the Saturn V," he said. "It stands higher than the Statue of Liberty. Now bring such a monster in the air in stable flight! The very first Saturn V went up like a candle. And why? Because of our alleged 'over-engineering.' "[138] In von Braun's interview with *Time* for the February 17, 1958, issue, he linked the rocket team's cohesion and self-reliance to the successful launch of Explorer 1, America's answer to Sputnik: "What corporation would have sent up a satellite two weeks ago?"[139] Of course, several Paperclippers ultimately joined corporations, including Dornberger. The "rubbing together" that Huzel lauded exemplified the MIC at an early stage.

Wernher von Braun's reputation as a brilliant scientist and project manager is certainly based in fact, but it also benefits from the "Huntsville School" of historians and space program veterans who depict von Braun as a visionary. Von Braun's obsession with space exploration required accommodation with both the Wehrmacht and the US Army. He excelled at promoting himself, his team, and a space program capable of both expanding humanity's horizons and guaranteeing US military supremacy over a technologically advanced foe. Frederick Ordway and Mitchell Sharpe describe a conversation between von Braun and his associate Adolf Thiel, shortly before the move to Redstone Arsenal, in which von Braun expressed frustration with his superiors' shortsightedness: "We can dream about rockets to the Moon until Hell freezes over. Unless the people understand it and the man who pays the bill is behind it, no dice. You worry about your damned calculations, and I'll talk to the people."[140] A slightly romanticized version of events, perhaps, but von Braun enthusiastically performed the role of the "public scientist" to secure broad support and funding.[141] While never named in the landmark studies of C. Wright Mills and Harold Lasswell, the sociologists anticipated men like von Braun ascending to influential positions in the MIC. Lasswell's "specialists in violence" were no longer soldiers but civilian technocrats capable of managing sprawling enterprises.[142] Mills observed how "the warlords" belonging to the power elite increasingly relied on "public relations," noting that "they have spent millions of dollars and they have employed thousands of skilled publicists, in and out of uniform, in order to sell their ideas and themselves to the public and to the Congress."[143] Von Braun pursued every avenue and medium at his disposal, but he never lost sight of the military potential inherent in rocketry.

The Paperclippers were brought to the United States to build weapons, and von Braun sold the military his vision for a weaponized space station as skillfully as he romanticized space travel in magazines and in Walt Disney's "Man in Space" short film. In von Braun's 1947 novel, *The Mars Project*, he militarized space by including a space station, called Lunetta, capable of destroying Soviet military and industrial targets with nuclear missiles.[144] Von Braun was on familiar ground promoting the "ultimate weapon" to American audiences invested in national security. In a September 1952 speech to the Business Advisory Council for the Department of Commerce, von Braun argued for a new ultimate weapon leading to "a permanent peace." Rocketry, von Braun maintained, "is . . . capable of solving the world's peace problems more effectively than any other branch of science and engineering, and simultaneously—that is to say without additional expenditure—doing a great deal of advancement for mankind." The blending of military and humanitarian goals typified von Braun's approach to public relations. He worried about Soviet advances, especially since Germans under his tutelage at Peenemünde worked on parallel projects in Russia. He concluded the speech to business leaders by warning that the Soviets were close: "If we do not wish them to wrest the control of space from us, it's time, and high time we

Walt Disney, *left,* and Wernher von Braun, posing with a model V-2 rocket, January 1, 1954. Von Braun was a technical director on three Disney films for television about space exploration. Disney's enormously popular "Man in Space" episode of *Disneyland* featured von Braun and several other Paperclippers, further normalizing the idea that the rocket team had become "our Germans." *Source:* Great Images in NASA GPN-2000-000060.

acted!"[145] In the same year, von Braun delivered a speech entitled "Space Superiority as a Means for Achieving World Peace," in which he requested a four-billion-dollar budget and a ten-year commitment to enforce, in Michael Neufeld's words, "a Pax Americana on Earth."[146]

Von Braun invoked the Soviet threat before Sputnik, which only elevated the rocket team for its foresight and vindicated Paperclip for bringing the team to the United States. In his 1952 article, "Why I Chose America," von Braun portrayed the rocket team as loyal and happy Americans who preferred "hominy grits to sauerkraut and whiskey to schnapps." He also invoked Jesus Christ multiple times. Von Braun and many other Paperclippers became born-again Christians after moving to Alabama.[147] Von Braun condemned the Nazi regime he had once served (under duress, he would have readers believe) and drew explicit comparisons to the Soviet Union, the godless Communists who now occupied his homeland.[148] A decade later he spoke to the Huntsville Ministerial Association with the passion of a new convert, stating, "Progress is claimed by the Soviets on a purely materialistic basis, with sole reliance upon man's strength and ingenuity. Such spiritual poverty is pathetic."[149] Sputnik obviated the need to warn the American people about Soviet ingenuity, but von Braun did so consistently. In a January 1959 speech to the Associated General Contractors of America, von Braun praised the Soviets' "massive educational program designed to provide a reservoir of scientific and engineering talent."[150] A month later he told the University of Florida that "if we do not match the ambitious Communist intentions to visit the Moon with an equally determined US space flight program . . . we may in the not-too-distant future be surrounded by several planets flying the Hammer and Sickle flag."[151] As an ambitious new citizen who believed the Paperclippers' story was part of America's, von Braun instinctively appealed to Americans' anticommunism, Christian idealism, and fascination with conquering "vast new frontiers."[152]

Wernher von Braun not only served the MIC with distinction, he was one of its architects and the embodiment of the public scientist in the national security age. Collectively, the Paperclippers integrated into the emerging MIC with relative ease because of their past experience serving a totalitarian technocracy. Moreover, most Paperclip scientists, engineers, and family members assimilated into a forgiving American society. The national security state evolved in the shadow of the Cold War, and Paperclip, an unusual and risky program when first introduced, was redeemed by the cascade of crises in Europe and Asia, and the concomitant perception of American weakness. Paperclip had many stakeholders, including the armed services, corporations, universities, and opportunistic politicians. While the Paperclippers thrived as gainfully employed Americans, in both civil service and private industry, Project Paperclip was remembered as a brilliant decision. Public memory celebrated "Our Germans," von Braun especially, but the Paperclippers understood better than most Americans that all glory is fleeting.

Epilogue

Gather 'round while I sing you of Wernher von Braun,
A man whose allegiance
Is ruled by expedience.
Call him a Nazi, he won't even frown.
"Ha, Nazi sch-mazi," says Wernher von Braun.

Don't say that he's hypocritical,
Say rather that he's apolitical.
"Once the rockets are up, who cares where they come down?
That's not my department," says Wernher von Braun.

Tom Lehrer, "Wernher von Braun"

The Soviet Union launched Sputnik on October 4, 1957. The national security state seemed both terrified and energized by what Senator Lyndon Johnson called a "disaster . . . comparable to Pearl Harbor."[1] The news broke the same day that the recently confirmed secretary of defense, Neil McElroy, was visiting the ABMA in Huntsville. Walter McDougall describes the moment vividly:

> There was an instance of stunned silence. Then von Braun started to talk as if he had been vaccinated with a Victrola needle. In his driving urgency to unburden his feelings, the worlds tumbled over one another. "We knew they were going to do it! . . . We have the hardware on the shelf. For God's sake, turn us loose and let us do something. We can put up a satellite in sixty days, Mr. McElroy! Just give us the green light and sixty days!"

ABMA commander General John Medaris pushed back, "No, Wernher, ninety days." Medaris consoled McElroy, whose head was spinning from the day's events: "When you get back to Washington and all Hell breaks loose, tell them we've got the hardware down there to put up a satellite any time."[2] The rocket team delivered with Explorer 1 on January 30, 1958, launching von Braun to even greater levels of fame and influence in the MIC. The political fallout over Sputnik placed von Braun's rocket team at the center of a high-stakes rivalry between the armed services and an Eisenhower White House anxious to rein in exploding military budgets. Whether the response would be civilian or military Sputnik "invited another American lurch toward technocracy."[3] The Paperclippers once more capitalized on the crisis atmosphere and enjoined the race to the moon, confident in their influence and longevity. The *New York Times* reported on July 17, 1969, just days before *Apollo 11* made history, that 85 of the original 119 rocket team members still worked in the MIC.[4] As of April 1951, Paperclip had recruited 528 scientists and engineers, 459 of whom eventually immigrated to the United States.[5] The public fawning over the space program made one point clear—our Germans were better than their Germans. Indeed, Arthur Rudolph, still reeling from his "voluntary" denaturalization proceeding when interviewed for *Frontline* in 1987, called the moon landing "a German victory."[6]

As the Delphic oracle of space and champion of the ICBM, von Braun waded carefully into the political firestorm ignited by Sputnik.[7] Intent on reminding his superiors just how right he was, about Soviet capabilities and the lack of US preparedness, von Braun nonetheless avoided alienating "pro-space" Democrats in Congress. He left the politicking to the parties and commentators, most of whom rallied around the rocket team. Republican congressman Elford Cederberg, echoing the JIOA's assault on Paperclip's naysayers, wrote a scathing editorial in July 1958, blaming Truman for tying the hands of the few German scientists he had allowed into the country. Cederberg described an early ICBM designed by von Braun during World War II, but claimed, "The men could not build it simply because Truman in his folly stopped all long-range missile projects."[8] Other post-Sputnik profiles similarly portrayed von Braun as a frustrated genius fighting a short-sighted US bureaucracy. *Time*'s cover story of von Braun related the rocket team's struggles within the ABMA and with both Truman's and Eisenhower's administrations, noting, with the benefit of hindsight, naturally, that the talented German cadre had been ready to build Sputnik before Sputnik. "This is not a design contest," von Braun said. "It is a contest to get a satellite into orbit, and we're way ahead on this."

After being directed in 1955 to separate satellite development from weaponry and pursue a "dignified" project, von Braun, according to *Time*, "snorted": "I'm all for dignity. But this is a Cold War tool. How dignified would our position really be if a man-made star of unknown origin suddenly appeared in our skies?"[9] The Paperclipper was a prophet, and the public heaped praise on the Germans responsible for answering the Soviet threat despite a decade of supposed interference and underappreciation.

President Eisenhower respected von Braun's talents as a weapons designer, especially since he was on the receiving end of his deadly creations while liberating the continent as Supreme Allied Commander, but the cautious president decided to uproot the rocket team from its army home and embed it in a new entity. Eisenhower introduced NASA in an address to Congress on April 2, 1958, arguing that "a civilian setting" for exploring space "would emphasize the concern for our nation that outer space be devoted to peaceful and scientific purposes."[10] The president's disdain for militarization informed his decision to create NASA. Keith Glennan, NASA's first administrator, recalled his first meeting with the secretary of the army in 1958: "I had not realized how much of a pet of the Army's von Braun and his operation had become. He was its one avenue to fame in the space business . . . I finally left with my tail between my legs."[11] The Germans surely experienced déjà vu as the military sparred with a newly created government corporation over their fate. In 1944 the Wehrmacht and Albert Speer's organization eluded Heinrich Himmler using a similar method.[12] Von Braun was apprehensive about his future, even as his fame grew, and the civilians at NASA likewise distrusted the vaunted Paperclippers. Von Braun approached Glennan on the eve of the merger, seeking both clarity and a privileged position for himself and his team, some of whom had been with him for nearly a quarter of a century. Glennan recalled the heated exchange in his memoirs: "Wernher finally ended the conversation by saying, 'Look, all we want is a very rich and very benevolent uncle.' What a personality!"[13] Von Braun received his funding and autonomy over "launch vehicle systems," but his ambition hit a ceiling. "There was always a lingering resentment at the Washington end toward von Braun and his team," former White House staffer Charles Sheldon shared. "There were always rumors that von Braun would someday be head of NASA. But there was great sensitivity in Washington about racial and ethnic interests. . . . Von Braun would never be given a political position."[14]

The final years of the 1950s were heady days for the rocket team, and Hollywood rushed to exploit the inherently provocative story of the "Missile-man" von Braun. He was already a celebrity thanks to Walt Disney's "Man in Space" television episode, which aired in 1955. The extremely popular program, which included Eisenhower among its fans, featured von Braun and other members of the rocket team detailing their fantastic visions for space travel accompanied by Disney's trademark animation. "Man in Space" even inspired representatives from the Soviet space program to request a copy of the short film to show their audiences.[15] Historian De Witt Douglas Kilgore notes that von Braun personally benefited in the long term from "the Disney stamp of unimpeachable Americanism." Furthermore, Kilgore observed, "In a culture that celebrated the adventurous white masculinity exemplified by Theodore Roosevelt and Ernest Hemingway, von Braun easily fit into the mold of boy's idol."[16] The Disney collaboration sidestepped any mention of von Braun's biography. In contrast, the 1960 theatrical release of *I Aim at the Stars*, starring Curd Jürgens as von Braun, depicted America's most famous German immigrant, next to Albert Einstein perhaps, as the idealistic space enthusiast who had toiled listlessly for the Third Reich before crossing the Atlantic to realize his dreams. People as diverse as Keith Glennan and Dora concentration camp survivor Jean Michel saw through the facile portrait. Glennan, a great admirer of von Braun, attended a special screening and, according to his diary, "reacted rather negatively to the whole thing. Von Braun is made out be an anti-Nazi and seems to epitomize the scientist's lack of responsibility for the end use of the products of his mind."[17] Michel considered the film "harmful because it is cunningly made. But it is full of lies, passing straight from the bombing of Peenemünde to the Liberation, as if the interlude of Dora had never existed."[18] Satirist Mort Sahl is credited with suggesting a different title for von Braun's sanitized biopic: *I Aim at the Stars—but Sometimes I Hit London*.[19] Critics and comics aside, it would be another decade before the celebrated Paperclippers confronted the complex legacy that they and their American proponents had so steadfastly evaded.

Michael Neufeld identified three aspects of the early historiography regarding the German rocket teams: A romanticization of Peenemünde as "fundamentally aimed at space travel, rather than weapons development for Hitler"; the portrayal of German scientists as apolitical space enthusiasts; and the suppression of "almost all information about concentration camp labor and membership in Nazi organizations," specifically SS affiliations.[20] This conscious construction of a palatable public image for America's new heroes required ignoring a wealth of verifiable revelations about the atrocities committed at Mittelbau-Dora, protests and inflammatory memoirs by French and other

West European survivors, and war crime trials in both West and East Germany.[21] The "conspiracy of silence" involving von Braun, other Paperclippers, and the US government began unraveling after each of the foundational myths came under scrutiny, sometimes from unlikely sources.[22]

East Germany initiated a targeted campaign to discredit the rocket team and von Braun personally soon after the creation of NASA. In 1963, Julius Mader, an East German lawyer and journalist who frequently worked for the Stasi (Ministry of State Security) published *Secret of Huntsville: The True Career of Rocket Baron Wernher von Braun*. Although the book is replete with florid language and overt propaganda, Julius Mader's research proved accurate, as evidenced by subsequent scholarship and the fact that Western intelligence agencies had the same information.[23] Mader, according to the East German press release, "investigated the past of the aristocrat in SS uniform and found numerous hitherto unknown facts about the former top secret fascist rocket armament."[24] NASA's director of security wrote J. Edgar Hoover that the book "should not be accepted without rebuttal" and correctly identified Mader's campaign as "part of the political scheme of the Soviets to prevent the US (and the free world) from being the first on the Moon."[25] Anxious to press the issue further, East Germany's official film studio DEFA produced *Frozen Lighting* in 1967, using Mader's book as the principal source. The film, which screened in Western Europe after editing out the anti–von Braun content, was one of the most expensive East German films ever made.[26] With most of the rocket team's original personnel files stored in East German archives, it was only a matter of time before unpleasant truths were revealed to a less skeptical audience.

Dismissing Communist propaganda was one thing, but von Braun found it difficult to avoid the Mittelbau-Dora trials held in Essen between 1967 and 1970. The US Army's Dachau trial in 1947 implicated Georg Rickhey, although he was acquitted, but no other Paperclipper faced prosecution, despite a significant paper trail connecting them to war crimes. The East German prosecutor in Essen could not summon von Braun to East Germany to act as a witness, but von Braun reluctantly agreed to give a deposition in June 1969 (just one month before the *Apollo* moon landing) in connection to the trial of three former SS men. NASA general counsel Paul Dembling recalled that von Braun seemed worried and was reluctant to return to Germany, afraid "they were going to do something to him."[27] NASA shielded its German employees as best it could and managed to minimize press coverage of the unusual episode, at least temporarily.[28] During his deposition, von Braun finally admitted witnessing the horrific conditions inside the tunnels, but he explicitly lied about the presence of slave labor at Peenemünde, knowing full well his own complicity in the matter.[29] In 1976, shortly before his death, von Braun told an interviewer he had seen "Mittelwerk several times, once while these prisoners were blasting new tunnels and it was a pretty hellish environment. I'd never been in a mine before, but it was clearly worse than a mine."[30] Michael Neufeld rightfully notes that while von Braun might have feared protesting worker conditions or using slave labor to his superiors, something he claimed early on, "Holocaust scholars have shown that almost no one was punished for excusing themselves from much more serious crimes."[31] Wernher von Braun died at age sixty-five from pancreatic cancer on June 16, 1977. Lauded as "the most visible symbol of the Space Race with the Soviets" and the "vindicated prophet of spaceflight," von Braun died nearly a decade before journalists and his own government exhumed the mountain of "revised" files and evidence that cast a harsh light on both Project Paperclip and the Paperclippers, most of whom had retired from their influential positions in the MIC.[32]

The Arthur Rudolph saga, which played out in the 1980s, did not destroy Paperclip's formidable legacy in the public imagination so much as revive dormant debates over Paperclip's efficacy and morality. The establishment of the OSI in 1979 meant disclosures relating to Paperclippers' pasts—from scholars, journalists, and most importantly, survivors—could actually find traction in the right hands. Residing within the Justice Department, the OSI owed its existence to Brooklyn congresswoman Elizabeth Holtzman's effort to create an independent office charged with prosecuting war criminals residing in the United States. Frequent revelations of "the Nazi next door" prompted the legislation, and while the OSI officially targeted violations of international law and crimes against humanity, hunting Nazis who entered the United States, either illegally or with help from other government entities, consumed the OSI's agenda. OSI investigator Eli Rosenbaum stumbled onto Rudolph's complicity in organizing and exploiting slave labor, specifically Rudolph's involvement in at least one mass hanging right outside his office window, after reading Frederick Ordway and Mitchell Sharpe's *The Rocket Team* and Jean Michel's memoir of Dora between 1979 and 1980.[33] OSI chief Neal Sher balked initially, noting "those Paperclip cases don't go anywhere." Sher also cautioned Rosenbaum that most Americans believed the von Braun team were "the good Nazis."[34] Nonetheless, Sher encouraged him to build the complex case, knowing full well the political ramifications of taking on the retired NASA engineer, the "Father of the Saturn Rocket."[35]

Arthur Rudolph, with the top half of a model Saturn V rocket. Rudolph, the Saturn V project manager at NASA Marshall Space Flight Center, in Huntsville, Alabama, voluntarily renounced his US citizenship in 1984 rather than face prosecution for war crimes committed while operations director for the V-2 factory in Mittelwerk. *Source:* NASA.

The case against Rudolph revealed both Rudolph's and the national security bureaucracy's deception about his role as head of Mittelwerk's Technical Division. No one doubted Rudolph was an "ardent Nazi," a characterization even the JIOA had to acknowledge; but whether he had been directly complicit in war crimes had divided investigators over the years. In his 1947 security interview, Rudolph admitted attending the hanging of Dora prisoners accused of sabotage and ordering laborers under his control to "bear witness." Rudolph also received daily prisoner strength numbers and was briefed on "new arrivals."[36] Herschal Auerbach, a US Army war crimes investigator involved in Rudolph's 1947 trial, recalled the absence of "damning evidence" at the time. There was a lack of coordination, he told Linda Hunt later, and "[e]vidence from prisoners was limited."[37] Eli Rosenbaum and Neal Sher began their investigation with considerably more information at their disposal and interviewed Rudolph on two separate occasions without the presence of a lawyer. The transcripts from the 1982 and 1983 interviews are compelling because Rosenbaum and Sher sat with Rudolph and revisited the original 1947 interview line by line. Rudolph eventually admitted knowing more than he had initially stated and divulged incriminating details in the 1980s interviews.[38] The OSI accused Rudolph of persecuting slave labor and engaging in a form of "terror." Rudolph's entry had violated State Department regulations, the OSI argued, and he clearly "lacked the good moral character essential for citizenship."[39] Nearly four decades later, Samuel Klaus's objections were finally being heard and acted on. The OSI and Arthur Rudolph agreed a trial was in neither party's interest. Rudolph voluntarily forfeited his US citizenship in 1984 and returned to Hamburg, where he continued to receive his US government pension until his death in 1996. Rudolph was the only Paperclip scientist the OSI investigated and determined prosecutable.[40]

Arthur Rudolph defended his legacy in "exile" while his fellow Paperclippers organized a spirited defense of one of their own, perhaps fearing their time would come. Several Paperclippers stressed the "Jewishness" of Rudolph's detractors and noted that most of the criticism of the German presence in the MIC, over forty years' worth, derived from Jews.[41] Politicians from every branch of government, in both parties, along with military and intelligence community personnel denounced Rudolph's treatment and the OSI's unfair "persecution" of an American citizen. Historians, journalists, Lyndon LaRouche, Holocaust deniers, and others joined the fray as well. Frederick Ordway, for example, whose own book had inspired Eli Rosenbaum to dig deeper, argued in 1988 that the Rudolph case be reopened. Ordway inexplicably cited Rudolph's sanitized 1946 OMGUS dossier to discredit the OSI's more thorough investigations.[42] James Murphy, Rudolph's deputy at NASA between 1965 and 1969, spoke for many Rudolph supporters in the United States when he declared himself "ashamed, very ashamed, that our country would have taken a man. Regardless of what may have happened before he came here, they had accepted him here. He had not only done his job, but he had really, in my opinion, done so much toward contributing to the wealth of this country."[43] Ed Buckbee, the director of the Space and Rocket Center at the time, wondered why it was acceptable to use Rudolph for forty years knowing his background and then punish him after he retired.[44] The city council of Huntsville, home first to the ABMA and later NASA's Marshall Space Flight Center, unanimously voted to defend Rudolph and to encourage President Ronald Reagan and Congress to somehow rectify things. Councilwoman Jane Mabry acknowledged that her German constituents had pressured the city to respond, but most citizens agreed the rocket team had enriched Huntsville in more ways than

one.[45] Councilman James Wall echoed Mabry's perspective, noting Rudolph "may have made a mistake, and you can't judge him based on the past. Judge him on the whole life."[46]

The Rudolph episode demonstrated just how integrated and assimilated "Our Germans" had become after decades of living as naturalized American citizens. On the other hand, the affair underscored lingering discomfort with the transactional nature of a program like Paperclip in those crucial early years of the Cold War. The decision to protect, employ, and harbor former Nazis because of what they offered the national security state predictably underwent greater scrutiny with each new generation. The national security state reflects what William Appleman Williams described as "a steady, if unacknowledged drift toward militarization." However, he adds, "it was the civilians who defined the world in military terms, not the military who usurped civilian power."[47] If one could write an epitaph for Project Paperclip and the national security state that created it, one could scarcely do better than the *Washington Star*'s editorial on the occasion of Wernher von Braun's death: "A kind of Faustian shadow may be discerned in—or imposed on—the fascinating career of Wernher von Braun: A man so possessed of a vision, of an intellectual hunger, that any accommodation may be justified in its pursuit."[48] Indeed, Project Paperclip is just one manifestation of the Faustian shadow that fell over the United States for the entirety of the Cold War. The United States came to embrace von Braun and all that he represented, a truth dramatized very effectively in the conclusion of *I Aim at the Stars*. Major Bill Taggert, a character who begins the film skeptical of the amoral scientist and contemptuous of von Braun personally, especially after losing his family in a V-2 attack on London, ultimately chooses to forgive the genius German scientist and acknowledge his idealism:

TAGGERT (putting on his coat): You know, I've almost grown to like you, von Braun.

VON BRAUN: Suppose I could say the same about you, except one never really likes one's conscience.

TAGGERT: But if I tried for a million years, I could never really understand you. . . . What have you scientists got in here, in the place of a sense of human values?

VON BRAUN: A concern for the future perhaps. The universe. The whole universe is waiting for us, and we must explore it. That's what makes man, man and not a mere vegetable.

TAGGERT: Goodbye, von Braun. And good luck with the universe.

The End[49]

Abbreviations

AAF	Army Air Forces
AFOSI	Air Force Office of Special Investigations
AGWAR	Adjutant General War Department
CG	commanding general
CIC	Counter-Intelligence Corps
CIOS	Combined Intelligence Objectives Subcommittee
CSGID	Office of the Chief of Staff, Ground Intelligence Division
EUCOM	European Command
FIAT	Field Intelligence Agency, Technical
FOIA	Freedom of Information Act
GSC	General Staff Corps
GSUSA	General Staff United States Army
IIR	Records of the Investigative Records Repository
IWG	Interagency Working Group
JCS	Joint Chiefs of Staff
JIOA	Joint Intelligence Objectives Agency
NASM	Smithsonian National Air and Space Museum Archives
NARA	National Archives and Records Administration
OMGUS	Office of Military Government, United States
RG	Record Group
SAC	special agent in charge
SCAP	Supreme Command for the Allied Powers
SHAEF	Supreme Headquarters Allied Expeditionary Force
SWNCC	State-War-Navy Coordinating Committee
USHMM	United States Holocaust Memorial Museum
USSAF	US Strategic Air Forces
WDGS	War Department General Staff
WFO	Washington Field Office

Introduction

1. Several members of the German rocket team claimed the German High Command had requisitioned a "death train" for rocket specialists near Regensburg. Scientists were to be "disposed of" for unsatisfactory work and to prevent Allied capture. See Agent report, Gröttrup Rocket Testing Institute, Moscow, July 16, 1948, folder 1, vol. 1, box 28, Records of the Investiga-

tive Records Repository (IIR), Counter-Intelligence Corps (CIC), Records of the Army Staff, Record Group (RG) 319, National Archives and Records Administration (NARA), College Park, Maryland; and Bob Ward, *Dr. Space: The Life of Wernher von Braun* (Annapolis, MD: Naval Institute Press, 2009), 55. Von Braun later claimed he learned that Hitler had instructed the SS to "gas all the technical men concerned with rocket development." See report of SA___, file 116-24800: Wernher Magnus Maximilian von Braun, May 22, 1960, Wernher von Braun, FBI Vault, http://vault.fbi.gov/Wernher%20VonBraun.

2. Richard Zoglin, "Bob Hope's 10 Best Jokes," *Vulture*, Nov. 14, 2014, http://www.vulture.com/2014/11/10-best-bob-hope-jokes.html.

3. Clarence Lasby, *Project Paperclip: German Scientists and the Cold War* (New York: Atheneum, 1971), 155.

4. Memo to all district directors from Argyle R. Mackey, commissioner, US Citizenship and Immigration Services, Dec. 8, 1952, box 38, Project Decimal File, 1951–1952, RG 319, Records of the Army Staff, NARA.

5. Burghard Ciesla, "German High Velocity Aerodynamics and Their Significance for the United States Air Force, 1945–1952," in *Technology Transfer out of Germany after 1945*, ed. Matthias Judt and Burghard Ciesla (Amsterdam: Harwood Academic Publishers, 1996), 93–106.

6. Walter A. McDougall, *The Heavens and the Earth: A Political History of the Space Age* (New York: Basic Books, 1985), 42.

7. Von Braun biographer Michael J. Neufeld emphasizes the continuities of the Nazi and American rocket program and the need to justify them in "Creating a Memory of the German Rocket Program for the Cold War," in *Remembering the Space Age*, ed. Steven J. Dick (Washington, DC: NASA, 2008), 71–87.

8. Linda Hunt's liberal use of the Freedom of Information Act (FOIA) and interviews with scientists, government officials, and victims of atrocities associated with Paperclip personnel resulted in significant revelations about government involvement in manipulating background investigations of Paperclip scientists. See *Secret Agenda: The US Government, Nazi Scientists, and Project Paperclip, 1945–1990* (New York: St. Martin's Press, 1991). Tom Bower, a British journalist, wrote *The Paperclip Conspiracy: The Hunt for the Nazi Scientists* (Boston: Little, Brown, 1987) and produced a documentary for PBS's *Frontline* entitled "The Nazi Connection," featuring interviews with subjects in his book. Annie Jacobsen's recent book is an exciting narrative replete with notorious Paperclippers' transgressions and further evidence of US government complicity. See *Operation Paperclip: The Secret Intelligence Program that Brought Nazi Scientists to America* (New York: Little, Brown, 2014). Eric Lichtblau devotes a chapter to Paperclip summarizing these findings in *The Nazis Next Door: How America Became a Safe Haven for Hitler's Men* (New York: Houghton Mifflin, 2014).

9. The IWG released approximately 8.5 million pages to the public between 1998 and 2007. Materials related directly to Paperclip include Office of Strategic Services (OSS) and CIA operational files, FBI case files, and twenty thousand pages from the CIC. See Nazi War Crimes and Japanese Imperial Government Records Interagency Working Group, *Final Report to the United States Congress, April 2007* (College Park, MD: IWG, 2007), http://www.archives.gov/iwg/reports/final-report-2007.pdf.

10. Gregory McLauchlan, "World War II and the Transformation of the US State: The Wartime Foundations of US Hegemony," *Sociological Inquiry* 67, no. 1 (Feb. 1997): 12.

11. Quoted in Daniel Yergin, *Shattered Peace: The Origins of the Cold War and the National Security State* (Boston: Houghton Mifflin, 1977), 194.

12. Quoted in Michael Sherry, *Preparing for the Next War: American Plans for Postwar Defense, 1941–1945* (New Haven, CT: Yale University Press, 1977), 24.

13. Yergin, *Shattered Peace*, 200.

14. Melvyn P. Leffler, "The American Conception of National Security and the Beginnings of the Cold War, 1945–48," *American Historical Review* 89, no. 2 (April 1984): 349.

15. Thomas H. Etzold, "Organization for National Security, 1945–1950," in *Containment: Documents on American Foreign Policy and Strategy, 1945–1950*, ed. Thomas H. Etzold and John Lewis Gaddis (New York: Columbia University Press, 1978), 8.

16. Sherry, *Preparing for the Next War*, ix.

17. Gregory Hooks and Gregory McLauchlan, "The Institutional Foundation of Warmaking: Three Eras of US Warmaking, 1939–1989," *Theory and Society* 21, no. 6 (Dec. 1992): 771.

18. Melvyn P. Leffler, *A Preponderance of Power: National Security, the Truman Administration, and the Cold War* (Stanford, CA: Stanford University Press, 1992), 41.

19. Etzold, "Organization for National Security," 22.

20. National Security Council, Executive Secretary, *United States Objectives and Programs for National Security: A Report to the National Security Council*, NSC 68 (Washington, DC: National Security Council, April 12, 1950), https://www.trumanlibrary.org/whistlestop/study_collections/coldwar/documents/pdf/10-1.pdf.

21. Steven Casey, "Selling NSC-68: The Truman Administration, Public Opinion, and the Politics of Mobilization, 1950–51," *Diplomatic History* 29, no. 4 (Sept. 2005): 690.

22. Sherry, *Preparing for the Next War*, 158.

23. Walter E. Grunden et al., "Laying the Foundation for Wartime Research: A Comparative Overview of Science Mobilization in National Socialist Germany, Japan, and the Soviet Union," *Osiris* 20 (2005): 80.

24. Quoted in Sherry, *Preparing for the Next War*, 216. Vannevar Bush's *Science: The Endless Frontier* (Washington, DC: GPO, 1945) convincingly detailed the benefits of investing in civilian and military science.

25. Naomi Oreskes, "Science and the Origins of the Cold War," in *Science and Technology in the Global Cold War*, ed. Naomi Oreskes and John Krige (Cambridge, MA: MIT Press, 2014), 16.

26. C. Wright Mills, *The Power Elite* (1956; repr., Oxford: Oxford University Press, 2000).

27. Ibid., 216–17.

28. De Witt Douglas Kilgore, "Engineers' Dreams: Wernher von Braun, Willy Ley, and Astrofuturism in the 1950s," *Canadian Review of American Studies* 27, no. 2 (1997): 108.

29. James Ledbetter, *Unwarranted Influence: Dwight D. Eisenhower and the Military-Industrial Complex* (New Haven, CT: Yale University Press, 2011), 6.

30. Farewell Address, Jan. 17, 1961, in Ledbetter, *Unwarranted Influence*, 217.

31. Zuoyue Wang, *In Sputnik's Shadow: The President's Science Advisory Committee and Cold War America* (New Brunswick, NJ: Rutgers University Press, 2009), 2.

32. Quoted in Richard V. Damms, "James Killian, the Technological Capabilities Panel, and the Emergence of President Eisenhower's 'Scientific-Technological Elite,'" *Diplomatic History* 24, no. 1 (Winter 2000): 65–66.

33. Ibid., 57–58.

34. Stuart W. Leslie, *The Cold War and American Science* (New York: Columbia University Press, 1993), 2.

35. Andrea J. Wolfe, *Competing with the Soviets: Science, Technology, and the State in Cold War America* (Baltimore, MD: Johns Hopkins University Press, 2013), 41.

36. Quoted in Helen Bury, *Eisenhower and the Cold War Arms Race: "Open Skies" and the Military-Industrial Complex* (London: I. B. Tauris, 2013), 4.

37. Quoted ibid., 24.

38. Harold D. Lasswell, *Essays on the Garrison State* (New Brunswick, NJ: Transaction Publishers, 1997), 56–59.

39. Quoted in Michael Bar-Zohar, *The Hunt for German Scientists*, trans. Len Ortzen (London: Arthur Baker, 1967), 11.

40. Michael J. Neufeld, "The Guided Missile and the Third Reich: Peenemünde and the Forging of a Technological Revolution," in *Science, Technology and National Socialism*, ed. Monika Renneberg and Mark Walker (Cambridge: Cambridge University Press, 1994), 51.

41. Ibid., 70.

42. Final report summary, folder 51, box 3, Field Intelligence Agency, Technical (FIAT), RG 260, Records of US Occupation Headquarters, World War II, NARA.

43. Andreas Heinemann-Grüder, "'Keinerlei Untergang': German Armaments Engineers during the Second World War and in the Service of the Victorious Powers," in Renneberg and Walker, *Science, Technology and National Socialism*, 50.

44. See Peter Baldwin, *Hitler, the Holocaust and the Historians Dispute* (Boston: Beacon Press, 1990).

45. Jean Michel, *Dora: The Nazi Concentration Camp Where Modern Space Technology Was Born and 30,000 Prisoners Died* (New York: Holt, Rinehart and Winston, 1979), 98.

46. Ibid., 2–3.

47. Ibid., 92.

48. Frederick I. Ordway III worked with Wernher von Braun at the Army Ballistic Missile Agency and NASA's Marshall Space Flight Center in Huntsville, Alabama, before devoting his life's work to valorizing the Paperclippers composing the rocket team. See Frederick I. Ordway III and Mitchell R. Sharpe, *The Rocket Team* (New York: Crowell, 1979; Burlington, Ontario: Apogee Books, 2003). Ordway became a vocal defender of Arthur Rudolph after his self-deportation in the wake of the Office of Special Investigations inquiry into his Nazi past. See Ordway, "Rudolph Case Should Be Reopened," *Aerospace America* 26 (Aug. 1988): B52.

49. Michael B. Petersen, *Missiles for the Fatherland: Peenemünde, National Socialism, and the V-2 Missile* (Cambridge: Cambridge University Press, 2009), 6.

50. Michael J. Neufeld, "Rolf Engel vs. the German Army: A Nazi Career in Rocketry and Repression," *History and Technology* 13, no. 1 (1996): 53. Neufeld helped initiate a full inquiry into the issue of slave labor at the Nordhausen Mittelwerk complex by revealing several memos implicating members of the rocket team, including Arthur Rudolph. I examine this in depth in chapters 1 and the epilogue. See Neufeld, "Wernher von Braun, the SS, and Concentration Camp Labor: Questions of Moral, Political, and Criminal Responsibility," *German Studies Review* 25, no. 1 (2002): 57–78.

51. Rip Bulkeley, *The Sputnik Crisis and Early United States Space Policy: A Critique of the Historiography of Space* (Bloomington: Indiana University Press, 1991), 204–5.

52. Roger D. Launius, "The Historical Dimension of Space Exploration: Reflections and Possibilities," *Space Policy* 16 (2000): 23–38.

53. Michael J. Neufeld, *Von Braun: Dreamer of Space, Engineer of War* (New York: Vintage Books, 2007), ix.

54. The city of Huntsville itself is a part of the historiography, given its well-deserved moniker Rocket City after the transfer of the rocket team to Redstone Arsenal. Monique Laney examines the assimilation of the Germans into a southern city in the throes of the civil rights movement in *German Rocketeers in the Heart of Dixie: Making Sense of the Nazi Past during the Civil Rights Era* (New Haven, CT: Yale University Press, 2015).

55. John Gimbel, "German Scientists, US Denazification Policy, and the 'Paperclip Conspiracy,'" *International History Review* 12, no. 3 (Aug. 1990): 442.

56. See John Gimbel, *Science, Technology, and Reparations: Exploitation and Plunder in Postwar Germany* (Stanford, CA: Stanford University Press, 1990); and Gimbel, "Project Paperclip: German Scientists, American Policy, and the Cold War," *Diplomatic History* 14, no. 3 (1990): 343–66. John Farquharson examined the British experience with German technology

and scientific personnel in "Governed or Exploited? The British Acquisition of German Technology, 1945–48," *Journal of Contemporary History* 32, no. 1 (Jan. 1997): 23–42.

57. Lasby, *Project Paperclip*, vii–ix.

58. Memo to A. H. Belmont from S. B. Donahoe, request received from Clarence G. Lasby of University of California, re Paper Clip Program, May 18, 1960, section 42, file 105-8090, box 92, RG 65, Records of the Federal Bureau of Investigation, NARA.

59. FBI director Hoover to Clarence Lasby, May 19, 1960, section 42, file 105-8090, box 92, RG 65, NARA.

60. William Appleman Williams, *The Tragedy of American Diplomacy* (1972; repr., New York: Norton, 2009), 291.

61. Judy Feigin, *The Office of Special Investigations: Striving for Accountability in the Aftermath of the Holocaust*, Dec. 2008 (Washington, DC: Department of Justice, 2010), https://www.justice.gov/sites/default/files/criminal/legacy/2011/03/14/12-2008osu-accountability.pdf.

Chapter 1 · Aristocracy of Evil

1. "Walter Jessel Spent Life Fighting for Change," *Daily Camera*, April 15, 2008, http://www.dailycamera.com/ci_13140551.

2. Grunden et al., "Laying the Foundation," 86.

3. Joseph Haberer, *Politics and the Community of Science* (New York: Van Nostrand Reinhold, 1969), 104–7. See also Mark Walker, "The Nazification and Denazification of Physics," in *Technology Transfer out of Germany after 1945*, ed. Matthias Judt and Burghard Ciesla (Amsterdam: Harwood Academic Publishers, 1996), 51.

4. Petersen, *Missiles for the Fatherland*, 29.

5. William Sims Bainbridge, *The Spaceflight Revolution: A Sociological Study* (New York: Wiley, 1983), 50.

6. Heinemann-Grüder, "Keinerlei Untergang," 35–37.

7. John Cornwell, *Hitler's Scientists: Science, War, and the Devil's Impact* (New York: Penguin, 2003), 9. See Michael Wildt, *An Uncompromising Generation: The Nazi Leadership of the Reich Security Main Office*, trans. Tom Lampert (Madison: University of Wisconsin Press, 2010).

8. Alan D. Beyerchen, *Scientists under Hitler: Politics and the Physics Community in the Third Reich* (New Haven, CT: Yale University Press, 1977), 9.

9. Margit Szöllösi-Janze, "National Socialism and the Sciences: Reflections, Conclusions and Historical Perspectives," in *Science in the Third Reich*, ed. Margit Szöllösi-Janze (New York: Berg, 2001), 12.

10. Beyerchen, *Scientists under Hitler*, 40.

11. Ibid., 1.

12. Szöllösi-Janze, "National Socialism and the Sciences," 12.

13. Petersen, *Missiles for the Fatherland*, 3.

14. See Michael Geyer's brilliant essay "German Strategy in the Age of Machine Warfare, 1914–1945," in *Makers of Modern Strategy: From Machiavelli to the Nuclear Age*, ed. Peter Paret (New York: Oxford University Press, 1986), 537–97.

15. Monika Renneberg and Mark Walker, "Scientists, Engineers and National Socialism," in Renneberg and Walker, *Science, Technology and National Socialism*, 6.

16. See Ian Kershaw, *The Nazi Dictatorship: Problems and Perspectives of Interpretation* (London: Bloomsbury Academic, 2000).

17. Paul Maddrell, *Spying on Science: Western Intelligence in Divided Germany, 1945–1961* (New York: Oxford University Press, 2006), 4.

18. Quoted in Charles R. Allen Jr., "Hubertus Strughold, Nazi in USA: Atrocities in the Name of Medical Science," *Jewish Currents* 28 (Dec. 1974): 5.

19. "Technical Advances in German Weapon Development," Jan. 16, 1945, folder 29, 17-18, box 33, FIAT General Records, RG 260, NARA.

20. Bainbridge, *Spaceflight Revolution*, 61.

21. Frank H. Winter, "Foundations of Modern Rocketry: 1920s and 1930s," in *Blueprint for Space: Science Fiction to Science Fact*, ed. Frederick I. Ordway III and Randy Liebermann (Washington, DC: Smithsonian Institution Press, 1992), 100.

22. Neufeld, "Guided Missile and the Third Reich," 58.

23. Eberhard Rees, interview by Michael Neufeld, Nov. 8, 1989, transcript, Peenemünde Interviews Project, 1989–1990, Smithsonian National Air and Space Museum Archives (NASM), Washington, DC.

24. Arthur Rudolph, interview by Michael Neufeld, Aug. 4, 1989, transcript, Peenemünde Interviews Project, 1989–1990, NASM.

25. David H. DeVorkin, *Science with a Vengeance: How the Military Created the US Space Sciences after World War II* (New York: Springer-Verlag, 1993), 25–26.

26. Michael J. Neufeld, "Hitler, the V-2, and the Battle for Priority, 1939–1943," *Journal of Military History* 57, no. 3 (July 1993): 537.

27. Quoted in DeVorkin, *Science with a Vengeance*, 52.

28. Quoted in Neufeld, *Von Braun*, 4.

29. The V-2's twisted road through the bureaucratic battlefield, especially Dornberger's role, is detailed in Neufeld, "Hitler, the V-2."

30. Bainbridge, *Spaceflight Revolution*, 58.

31. Gerhard Reisig, interview by Michael Neufeld, June 5–7, 1989, transcript, Peenemünde Interviews Project, 1989–1990, NASM.

32. Rees interview, Nov. 8, 1989, NASM.

33. Michael J. Neufeld, *The Rocket and the Reich: Peenemünde and the Coming of the Ballistic Missile Era* (Cambridge, MA: Harvard University Press, 1995), 192.

34. Samuel A. Goudsmit, *Alsos* (Los Angeles: Tomash, 1986), 145–46.

35. Wernher von Braun, "Survey of Development of Liquid Rockets in Germany and Their Future Prospects," *Journal of the British Interplanetary Society* 10, no. 2 (March 1951): 75.

36. Walter Dornberger, *V-2: The Nazi Rocket Weapon* (New York: Ballantine, 1954), 68.

37. Michel, *Dora*, 2–3.

38. A small concentration camp on the island of Usedom served the Peenemünde complex.

39. David Irving, *The Mare's Nest* (London: William Kimber, 1964), 309.

40. Basil Collier, *The Battle of the V-Weapons, 1944–1945* (Yorkshire: Emfield Press, 1976), 122.

41. Adam Tooze, *The Wages of Destruction: The Making and Breaking of the Nazi Economy* (New York: Penguin, 2006), 622–23.

42. Michael J. Neufeld, "Mittelbau Main Camp (aka Dora)," in *United States Holocaust Memorial Museum Encyclopedia of Camps and Ghettos, 1933–1945*, ed. Geoffrey P. Megargee (Bloomington: Indiana University Press, 2009), 966.

43. HQ Sub-Region Kassel, CIC Region III, Lager Dora, Jan. 31, 1947, folder 2, vol. 1, box 28, Records of the IIR, CIC, RG 319, NARA.

44. Gretchen Schafft and Gerhard Zeidler, *Commemorating Hell: The Public Memory of Mittelbau-Dora* (Urbana: University of Illinois Press, 2012), ix.

45. Andrè Sellier, *A History of the Dora Camp*, trans. Stephen Wright and Susan Taponier (Chicago: Ivan R. Dee, 2003), 5.

46. "Report: Hitler Tested V-2 Rockets by Firing Them at His Own People," *Haaretz*, March 13, 2015, http://www.haaretz.com/jewish/news/1.646611.

47. Gerald Reitlinger, *The SS: Alibi of a Nation, 1922–1945* (New York: Da Capo, 1989). Most of the Peenemünde personnel "made the SS a scapegoat for all the crimes associated with the

rocket program," even when some of these crimes were initiated by the civilian scientists. Neufeld, *Rocket and the Reich*, 184.

48. Neufeld, *Von Braun*, 159.

49. Petersen, *Missiles for the Fatherland*, 167.

50. Quoted in Jacobsen, *Operation Paperclip*, 13.

51. Neufeld, "Guided Missile and the Third Reich," 66.

52. A 2014 German television documentary claims Kammler's suicide was manufactured by American intelligence operatives to secure his knowledge of secret weapons programs. The sources of the claim are former Kammler associates, but most historians refute the claim that American intelligence officers staged his suicide and provided him a new identity. Given his reputation among the Paperclippers, few would have protected this secret indefinitely. See "Did US Fake Top Nazi's WWII Suicide and Spirit Him Away to Get Hands on Hitler's Secret Weapons Programme?," *Daily Mail*, June 10, 2014.

53. Reisig interview, June 5–7, 1989, NASM.

54. Oral history interview with Herbert Axster, RG-50.702*0001, Oral History Interviews of the Linda Hunt Collection, United States Holocaust Memorial Museum (USHMM).

55. Rees interview, Nov. 8, 1989, NASM.

56. Michael McDonald and Viorel Badescu, eds., *The International Handbook of Space Technology* (Berlin: Springer-Verlag, 2014), 4.

57. Oral history interview with Hugh Carey, RG-50.702*0024, Hunt Collection, USHMM. David Malachowsky also participated in the liberation of Nordhausen and similarly described bodies "stacked up like cordwood. Neatly." See oral history interview with David Malachowsky, RG-50.702*0031, Hunt Collection, USHMM.

58. HQ Twelfth Army Group Office of the Judge Advocate, "Report of Investigation of Alleged War Crime," May 25, 1945, document no. 2222-PS, in Nazi Conspiracy and Aggression, vol. 4, the Avalon Project, Lillian Goldman Law Library, Yale Law School, New Haven, Connecticut, http://avalon.law.yale.edu/subject_menus/nca_v4menu.asp.

59. Yves Béon, *Planet Dora: A Memoir of the Holocaust and the Birth of the Space Age*, trans. Yves Béon and Richard L. Fague (Boulder, CO: Westview Press, 1997), 132.

60. Oral history interview with Yves Béon, RG-50.702*0004, Hunt Collection, USHMM. Erich Traub, also a laborer at Dora, confirms the constant spectacle of hangings. Oral history interview with Erich Traub, RG-50.702*0017, Hunt Collection, USHMM.

61. National Archives and Records Service, *United States Army Investigation and Trial Records of War Criminals*, USA vs. Kurt Andrae et al. *(and Related Cases) April 27, 1945–June 11, 1958*, pamphlet describing M1079 (Washington, DC: National Archives Microfilm Publications, 1981), accessed June 1, 2015, http://www.archives.gov/research/captured-german-records/microfilm/m1079.pdf.

62. Michel, *Dora*, 97.

63. Béon, *Planet Dora*, 24.

64. Peter P. Wegener, *The Peenemünde Wind Tunnels: A Memoir* (New Haven, CT: Yale University Press, 1996), 95.

65. Petersen, *Missiles for the Fatherland*, 177. Arthur Rudolph spoke of constant shortages and production problems, citing one instance in which the quota for V-2s doubled from 900 to 1,800 per month. See audio recording of Arthur Rudolph, RG-50.702*0019, Hunt Collection, USHMM.

66. Rudolph interview, Aug. 4, 1989, NASM.

67. Neufeld, *Rocket and the Reich*, 186.

68. Quoted in Neufeld, *Von Braun*, 143.

69. Quoted in Petersen, *Missiles for the Fatherland*, 149.

70. Sellier, *History of the Dora Camp*, 26.

71. "Transcript of US Army Interrogation of Arthur Rudolph in June 1947," appendix A in Thomas Franklin, *An American Exile: The Story of Arthur Rudolph* (Huntsville, AL: Hugh McInnish, 1987), 172–74.

72. Ibid., 184–85.

73. Rudolph interview, Aug. 4, 1989, NASM.

74. Neufeld, *Rocket and the Reich*, 188.

75. Neufeld, *Von Braun*, 63.

76. Audio recording of Wernher von Braun, RG-50.702*0010, Hunt Collection, USHMM.

77. Neufeld, "Wernher von Braun, the SS," 63.

78. FBI file 77-594, Wernher Magnus Maximilian Freiherr von Braun, Sept. 15, 1948, SA James P. O'Neil, box 151, 105-10747, RG 65, NARA.

79. Neufeld, "Wernher von Braun, the SS," 64.

80. Ernst Stuhlinger and Michael J. Neufeld, "Wernher von Braun and Concentration Camp Labor: An Exchange," *German Studies Review* 26, no. 1 (2003): 125.

81. Neufeld, "Wernher von Braun, the SS," 68–69.

82. Von Braun audio, RG-50.702*0010, Hunt Collection, USHMM; and Neufeld, "Wernher von Braun, the SS," 65.

83. Quoted in Ordway and Sharpe, *Rocket Team*, 59.

84. Quoted in Bower, *Paperclip Conspiracy*, 218.

85. Discussion with US, French, and British FIAT, folder 26, box 1, FIAT, RG 260, NARA.

86. Second preliminary report on Fritz Ter Meer, Aug. 1, 1945, folder 30, box 2, FIAT, RG 260, NARA.

87. Annie Jacobsen details US interest in IG Farben's chemical weapon program in *Operation Paperclip*, 142–65.

88. Memo to assistant chief of staff, G-2, US Forces European Theater (USFET), from FIAT, Otto Ambros, Sept. 18, 1945, folder 14, 17-23, box 5, FIAT, RG 260, NARA. See also Office of Assistant Chief of Staff, G-2, Poison Gas, abstract from notes on interrogations at Frankfurt (April 21–May 4, 1945), May 18, 1945, folder 27, 17-2, box 8, FIAT, RG 260, NARA.

89. Memo to assistant chief of staff, G-2, USFET, from Ambros, Sept. 18, 1945, RG 260, NARA.

90. Bower, *Paperclip Conspiracy*, 226–31.

91. Lichtblau, *Nazis Next Door*, 101–3. See also Allen, "Hubertus Strughold, Nazi in USA," 9.

92. Walter Jessel, "A Travelogue Through a Twentieth-Century Life: A Memoir" (unpublished manuscript, 1996, courtesy of Alfred Jessel), 140.

93. Ibid.

94. Dornberger was interned in a British POW camp for two years before being released in 1947 to work at Wright Field Air Force Base. See Hunt, *Secret Agenda*, 30.

95. Stenographic notes of an interrogation of PW GS/2379 Gen. Dornberger, Aug. 20, 1945, box 649, RG 165, Records of the War Department General and Special Staffs, NARA.

96. Jessel, "Travelogue Through a Twentieth-Century Life," 141.

97. Ibid.

98. Quoted in Bar-Zohar, *Hunt for German Scientists*, 176.

99. Appendix A, HQ Third US Army Intelligence Center, Interrogation Center, Special Screening Report, June 12, 1945, box 8, FIAT, RG 260, NARA.

100. Quoted in Bower, *Paperclip Conspiracy*, 129.

101. Hunt, *Secret Agenda*, 29.

102. Special Interrogation Report, Office of Military Government, United States (OMGUS) FIAT, "Evidence of a Conspiracy among Leading German 'Overcast' Personnel," box 8, FIAT, RG 260, NARA.

103. Ibid.

104. Quoted in Bainbridge, *Spaceflight Revolution*, 67.

105. Interrogation Center, Special Screening Report, June 12, 1945, box 8, FIAT, RG 260, NARA.

106. Jacobsen, *Operation Paperclip*, 41.

107. Memo to commanding general (CG), Army Air Forces (AAF), from US Strategic Air Forces (USSAF) in Europe, director of intelligence, "Recommendations for Organizing Reference Center for German Scientific and Technical Material in USSTAF," May 5, 1945, signed by Theodore von Karman, director, Scientific Adviser Groups, AAF, in *History of the Army Air Forces Participation in Project Paperclip*, microfilm A2055, Air Force Historical Research Agency, United States Air Force, Washington, DC.

108. "World War II: Operation Paperclip," Jewish Virtual Library, accessed June 10, 2015, https://www.jewishvirtuallibrary.org/jsource/ww2/OperationPaperclip.html. Linda Hunt writes that US Navy captain Ransom Davis decided to use the Osenberg list in consultation with the JCS. See Hunt, *Secret Agenda*, 32.

109. Burghard Ciesla, "Das 'Project Paperclip'—deutsche Naturwissenschaftler und Techniker in den USA (1946 bis 1952)," in *Historische DDR-Forschung: Aufsätze und Studien*, ed. Jürgen Kocka (Berlin: Akademie Verlag, 1993), 294.

110. Ibid., 297.

111. Ibid., 295.

112. Bainbridge, *Spaceflight Revolution*, 58.

113. JIOA memo for Lt. Gen. S. J. Chamberlain, director of intelligence, General Staff United States Army (GSUSA), "Delays Encountered in the Department of Justice in Processing the Immigration into the US of German Scientists," April 27, 1948, box 2, JIOA General Correspondence, RG 330, NARA.

114. Wegener, *Peenemünde Wind Tunnels*, 126.

115. Hunt, *Secret Agenda*, 112.

116. Gimbel, "German Scientists, US Denazification Policy," 434.

117. Memo to JCS, JIOA, from EUCOM, "Security Standards for Paperclip Personnel," May 7, 1948, box 16, JIOA General Correspondence, RG 330, NARA.

118. Petersen, *Missiles for the Fatherland*, 109.

119. Rees interview, Nov. 8, 1989, NASM.

120. Mr. A. Devitt Vanech, special assistant to the attorney general, to director, FBI, "German Specialists and Scientists in the US under the Protective Custody of the JIOA," Feb. 17, 1947, box 7, JIOA General Correspondence, RG 330, NARA.

121. Quoted in Hunt, *Secret Agenda*, 281.

122. Neufeld, *Von Braun*, 70.

123. Basic personnel record, Magnus von Braun, box 20, Foreign Scientist Case Files, RG 330, NARA.

124. Personal statement, June 2, 1947, Magnus von Braun, box 20, Foreign Scientist Case Files, RG 330, NARA.

125. Security report, March 31, 1948, Magnus von Braun, box 20, Foreign Scientist Case Files, RG 330, NARA.

126. Extract of data from FBI files as forwarded to JIOA by Mr. A. Devitt Vanech, Feb. 21, 1947, Magnus von Braun, box 20, Foreign Scientist Case Files, RG 330, NARA.

127. FBI file 77-594, von Braun, Sept. 25, 1948, SA James P. O'Neil, 105-10747, box 151, RG 65, NARA.

128. Amy Gerber's documentary about her grandfather, von Braun associate Edward Gerber, revolves around her attempt to locate a cache of documents buried in her family's property in Germany. See Amy Gerber, director, *My Grandfather Was a Nazi Scientist: Opa, Von Braun and Operation Paperclip* (Middleburg, VA: Flatcoat Films, 2010).

129. "Report on Operation Oberjoch of Project Abstract on Prof. Wernher von Braun's and Dr. Herbert Axster's Connection with Operation Oberjoch on the Possibility of Tracing the Missing Oberjoch and Bad Sachsa Documents through von Braun and Axster," Sept. 27, 1947, box 31, Records of the IIR, CIC, RG 319, NARA.

130. Quoted in Neufeld, *Von Braun*, 96.

131. Ibid., 96–98.

132. Copy of immigration and security dossier, Feb. 19, 1947, FBI file 77-594, von Braun, Sept. 25, 1948, SA James P. O'Neil, 105-10747, box 151, RG 65, NARA.

133. Neufeld, "Wernher von Braun, the SS," 73.

134. Quoted in Linda Hunt, "US Coverup of Nazi Scientists," *Bulletin of Atomic Scientists* 48, no. 7 (April 1985): 20.

135. Signed statement by Arthur Rudolph, May 12, 1969, Rudolph Air Force Office of Special Investigations (AFOSI), RG 60, General Records of the Department of Justice, NARA.

136. Rudolph interview, Aug. 4, 1989, NASM.

137. Quoted in Hunt, "US Coverup of Nazi Scientists," 20.

138. Report on Rudolph from Henry Grant, Region IV, 111th CIC Det., April 1, 1953, Rudolph AFOSI, RG 60, NARA.

139. Schafft and Zeidler, *Commemorating Hell*, 91.

140. Memo to director of intelligence, War Department General Staff (WDGS), attn: chief, Intelligence Group, for chief, Special Exploitation Branch, from Office of the Director of Intelligence, HQ EUCOM, Georg Rickhey, box 135, Foreign Scientist Case Files, RG 330, NARA.

141. Georg Rickhey to Mr. Peter Beasley, April 6, 1948, Georg Rickhey, box 135, Foreign Scientist Case Files, RG 330, NARA.

142. Memo for director of intelligence, WDGS, "Accusations against Mr. Patin and Dr. Rickhey," Georg Rickhey, box 135, Foreign Scientist Case Files, RG 330, NARA. Kurt Kettler, an engineer partly responsible for Mittelwerk, was replaced by Rickhey on April 13, 1944. Kettler remembered Rickhey, a known "political activist" dressed in an immaculate Nazi uniform. Not surprisingly, Rickhey worked well with the SS leadership. See Ordway and Sharpe, *Rocket Team*, 59.

143. Statement by Willy Geiger, Sept. 30, 1947, Hermann Nehlsen, box 118, Foreign Scientist Case Files, RG 330, NARA; Accusations against Mr. Patin and Dr. Rickhey, RG 330, NARA.

144. Security report, signed by D. L. Putt, Oct. 25, 1946, Albert Patin, box 123, Foreign Scientist Case Files, RG 330, NARA.

145. Security report, signed by H. M. McCoy, Sept. 27, 1948, Albert Patin, box 123, Foreign Scientist Case Files, RG 330, NARA.

146. "Nazis Sent to US as Technicians," *New York Times*, Jan. 4, 1947.

147. Message from War Department director of intelligence to EUCOM, Aug. 20, 1947, vol. 5, box 77, Records of the IIR, CIC, RG 319, NARA.

148. HQ EUCOM memo to HQ Department of the Army for Office of the Chief of Staff, Ground Intelligence Division (CSGID), Oct. 2, 1947, vol. 5, box 77, Records of the IIR, CIC, RG 319, NARA; "Report on Operation Oberjoch of Project Abstract," Sept. 27, 1947, RG 319, NARA.

149. Revised security report for Herbert F. Axster and revised security report for Ilse Axster, Herbert Axster, box 5, Foreign Scientists Case Files, RG 330, NARA.

150. Revised security report for Ilse Axster, RG 330, NARA.

151. Memo, "The Axster Couple," May 22, 1947, Herbert Axster, box 5, Foreign Scientists Case Files, RG 330, NARA.

152. Gerhard Weise statement, n. d., Herbert Axster, box 5, Foreign Scientists Case Files, RG 330, NARA.

153. Statement of Mrs. Ilse Axster, signed May 22, 1947, Herbert Axster, box 5, Foreign Scientists Case Files, RG 330, NARA.

154. Revised security report for Herbert F. Axster, RG 330, NARA.

155. Memo to HQ EUCOM from Intelligence Division, WDGS, box 6, JIOA General Correspondence, RG 330, NARA.

156. Quoted in Christopher Simpson, *Blowback: The First Full Account of America's Recruitment of Nazis, and Its Disastrous Effect on Our Domestic and Foreign Policy* (New York: Weidenfeld and Nicolson, 1988), 34.

Chapter 2 · Implements of Progress

1. Quoted in DeVorkin, *Science with a Vengeance*, 49.

2. Hitler's Scorched Earth Decree (Nero Decree), March 19, 1945, and Albert Speer's response, March 29, 1945, German History in Documents and Images, accessed June 18, 2015, http://germanhistorydocs.ghi-dc.org/sub_document.cfm?document_id=1590.

3. Gimbel, *Science, Technology, and Reparations*, 3.

4. Quoted in Jacobsen, *Operation Paperclip*, 11.

5. Paul Maddrell, "British-American Scientific Intelligence Collaboration during the Occupation of Germany," *Intelligence and National Security* 15, no. 2 (2000): 79.

6. Quoted in Sean Longden, *T-Force: The Forgotten Heroes of 1945* (London: Constable and Robinson, 2009), 86.

7. Joseph Mark Scalia, *Germany's Last Mission to Japan: The Failed Voyage of U-234* (Annapolis, MD: Naval Institute Press, 2000).

8. Carl Boyd and Akihiko Yoshida, *The Japanese Submarine Force and World War II* (Annapolis, MD: Naval Institute Press, 2002), 164.

9. Report on Natter Interceptor Project, from Clark S. Millikan, June 8, 1945, in United States Army, Forces in the Branch European Theatre, Research and Intelligence Branch, *Peenemünde East, Through the Eyes of 500 Detained at Garmish* (Washington, DC: US Army, ca. 1945), Internet Archive, https://archive.org/details/PeenemundeeasttooOMar.

10. US War Department, Military Intelligence Division, *German Technical Aid to Japan: A Survey* (Washington, DC: War Dept., Aug. 31, 1945), https://archive.org/details/German TechnicalAidToJapan.

11. Report on Emil M. Briner from Donald F. Jackson, 108th CIC Det., folder 4, box 27, Records of European Theater of Operations, US Army, RG 338, Records of US Army Operational, Tactical, and Support Organizations, NARA.

12. Memo to Adjutant General War Department (AGWAR) for Combined Chiefs of Staff from SHAEF Main, signed Eisenhower, n. d., box 6, Administrative History, FIAT, RG 260, NARA.

13. Memo from Office of Assistant Chief of Staff, G-2, to chief of staff, establishment of a FIAT, G-2, SHAEF, May 23, 1945, folder 54, 17-1, box 4, FIAT, RG 260, NARA.

14. Memo to chief, FIAT, deputy chief, FIAT, from K. W. J. Jones, FIAT, "Intelligence and Administrative Arrangements for Investigators," July 15, 1945, folder 17, 17-3, box 6, FIAT, RG 260, NARA.

15. "History of FIAT: Period 1 July 1946–30 June 1947," folder 28, box 1, FIAT, RG 260, NARA.

16. Final report summary, folder 51, box 3, FIAT, RG 260, NARA.

17. "Doc. 186: Use of German Technical Information, January 22, 1947," in *The Papers of General Lucius D. Clay: Germany 1945–1949*, vol. 1, ed. James Edward Smith (Bloomington: Indiana University Press, 1974), 305–6.

18. Quoted in Dik Alan Daso, "Operation LUSTY: The US Army Air Forces' Exploitation of the Luftwaffe's Secret Aeronautical Technology, 1944–45," *Aerospace Power Journal* 16 (Spring 2002): 31.

19. Memo to AGWAR for Combined Chiefs of Staff from Staff Main, signed Eisenhower, May 15, 1945, folder 44, box 3, FIAT, RG 260, NARA.

20. Memo to Berlin Office FIAT OMGUS from Sci.Br. FIAT, "Definition of the Term 'Scientist,'" March 30, 1946, folder 2, 17-4, box 7, FIAT, RG 260, NARA.

21. SHAEF (rear) report: progress report for 1945, June 4, 1945, folder 10, 17-3, box 5, FIAT, RG 260, NARA.

22. Memo to chief of staff, US Group CC, from Industry Division, Control Office (IG Farben), "Suggested Plan for Acquiring Technical 'Know How' in US Zone," Aug. 4, 1945, folder 19, 17-3, box 6, FIAT, RG 260, NARA.

23. Mitchell G. Ash, "Denazifying Scientists—and Science," in *Technology Transfer out of Germany after 1945*, ed. Matthias Judt and Burghard Ciesla (New York: Routledge Press, 2013), 62.

24. Quoted in John Gimbel, *The American Occupation of Germany: Politics and the Military, 1945–1949* (Stanford, CA: Stanford University Press, 1968), 1.

25. Memo to chief of staff from FIAT, "Movement of German Scientists and Technologists," signed by G. Brian Conrad, brigadier general, director of intelligence, Sept. 6, 1945, folder 1, 17-19, box 34, FIAT General Records, RG 260, NARA.

26. "Technical and Scientific Research in Germany after the War," June 1945, folder 26, 17-4, box 9, FIAT General Records, RG 260, NARA.

27. Ibid.

28. Ibid.

29. OMGUS FIAT to Colonel E. W. Gruhn, JIOA, Sept. 28, 1945, folder 25, 17-2, box 8, FIAT General Records, RG 260, NARA.

30. Quoted in Lasby, *Project Paperclip*, 58.

31. US Foreign Economic Administration, *Report on the Ad Hoc Interdepartmental (War-Navy) Committee to Handle Foreign Economic Administration Projects on Project 4: The Post-Surrender Treatment of German Engineering and Research in the "Secret Weapon" Field*, July 10, 1945 (Washington, DC: Foreign Economic Administration, 1945).

32. Ibid., 39–47.

33. Ibid., 51.

34. Ibid., 68.

35. Quoted in Wolfgang W. E. Samuel, *American Raiders: The Race to Capture the Luftwaffe's Secrets* (Jackson: University Press of Mississippi, 2004), 187.

36. Ibid., 68.

37. My interest concerns the JIOA's argumentation in favor of Paperclip and the method by which the agency manipulated the bureaucracy to expand Paperclip's mandate. For a detailed administrative history see Gimbel, "Project Paperclip," and *Science, Technology, and Reparations*.

38. James Forrestal to secretary of state James Byrnes, Jan. 29, 1945, box 6794, Central Decimal File, 862.542, RG 59, General Records of the Department of State, NARA.

39. Quoted in Lasby, *Project Paperclip*, 155.

40. Basic directive, JIOA, folder 17, 17-3, box 6, FIAT General Records, RG 260, NARA.

41. Lasby, *Project Paperclip*, 107.

42. Navy captain Bosquet Wev, an eventual JIOA director, was the agency's first deputy director. See memo for Major George Collins, executive director, CIC, July 18, 1945, box 1, JIOA General Correspondence, RG 330, NARA.

43. Quoted in Hunt, *Secret Agenda*, 10.

44. Quoted in Jacobsen, *Operation Paperclip*, 106.

45. Quoted in Lasby, *Project Paperclip*, 108–9.

46. David DeVorkin, "War Heads into Peace Heads: Holger N. Toftoy and the Public Image of the V-2 in the United States," *Journal of the British Interplanetary Society* 45 (1992): 439–41.

47. The complete report of Backfire is available in five volumes. See *Report on Operation Backfire*, 5 vols. (London: War Office, 1946), http://www.v2rocket.com/start/chapters/backfire.html.

48. Memo to director of intelligence from Ralph Osborne, chief, FIAT, "Policy Regarding V-Weapon Researchers," Aug. 6, 1945, folder 31-2, box 2, FIAT General Records, RG 260, NARA. See also Alan Beyerchen, "German Scientists and Research Institutions in Allied Occupation Policy," *History of Education Quarterly* 22, no. 3 (Autumn 1982): 292.

49. Memo to Mr. D. M. Ladd from E. G. Fitch, Navy Project 77, Aug. 8, 1945, 105-8090-8, box 45, RG 65, NARA.

50. Memo to Mr. E. G. Fitch from Mr. C. D. Marrow, Navy Project 77, Nov. 1, 1945, 105-8090-8, box 45, RG 65, NARA.

51. Memo to J. Edgar Hoover from special agent in charge (SAC), New York, Navy Project, Nov. 2, 1945, 105-8090-8, box 45, RG 65, NARA.

52. Scalia, *Germany's Last Mission to Japan*, 157–58.

53. Laney, *German Rocketeers*, 26–27.

54. Simpson, *Blowback*, 34.

55. Quoted in Gimbel, "Project Paperclip," 353; Gimbel, *Science, Technology, and Reparations*, 37.

56. Gimbel, "Project Paperclip," 345.

57. Bradley Dewey and Almy Chemical Company to Henry Wallace, Commerce, Oct. 23, 1945, box 6793, Central Decimal File, 862.542, RG 59, NARA.

58. Quoted in Gimbel, *Science, Technology, and Reparations*, 24–25.

59. Memo to President Harry Truman from secretary of commerce Henry Wallace, Nov. 29, 1945, box 1, JIOA General Correspondence, RG 330, NARA.

60. Memo for CG, Armed Service Forces, "Plan for Extended Exploitation of Overcast Personnel," from Clayton Bissell, major general, General Staff Corps (GSC), G-2, Feb. 6, 1946, box 4, JIOA General Correspondence, RG 330, NARA.

61. Memo to director, JIOA, from head of Technical Intelligence Center, draft of letter from the secretary of war to the secretary of state concerning exploitation of German Scientists for the benefit of US industry, comments on, box 4, JIOA General Correspondence, RG 330, NARA.

62. Robert Patterson, secretary of war, to secretary of state, Dec. 13, 1945, box 6793, Central Decimal File, 862.542, RG 59, NARA.

63. Secretary of war to secretary of state, n.d., box 4, JIOA General Correspondence, RG 330, NARA.

64. Quoted in Ciesla, "German High Velocity Aerodynamics," 101.

65. Quoted in Gimbel, "US Policy and German Scientists," 441.

66. Ciesla, "German High Velocity Aerodynamics," 99.

67. Memo, "Effects of Possibly Reduced Funds on Project Paperclip," May 1, 1947, signed H. M. McCoy, box 3693, RD 3723, RG 342, Records of the United States Air Force Commands, Activities, and Organizations, NARA.

68. Ash, "Denazifying Scientists," 61.

69. Aircraft Industries Association of America to General Henry H. Arnold, CG, AAF, Oct. 18, 1945, in *History of the Army Air Forces Participation in Project Paperclip* (May 1945–March 1947), part two, microfilm A2055, 791.

70. Memo from chief of naval operations to director, JIOA, Oct. 21, 1948, box 12, JIOA General Correspondence, RG 330, NARA.

71. Commerce memo, "Exploitation of German Scientists by the US Aviation Industry," Feb. 14, 1946, folders 1 and 2, box 42, part 1, State-War-Navy Coordinating Committee (SWNCC), 12, RG 353, Interdepartmental and Intradepartmental Committees, NARA.

72. Quoted in Sherry, *Preparing for the Next War*, 229.

73. Ibid.

74. Memo, "Effects of Possibly Reduced Funds on Project Paperclip," May 1, 1947, RG 342, NARA.

75. Ibid.

76. "Prevention of Scientists and Technicians from Leaving US Zones of Occupation in Germany and Austria," report by the SWNCC Subcommittee for Europe, folder 1, box 42, part 2, SWNCC, 257, 19-26, RG 353, NARA.

77. Memo for Col. Donald Putt from HQ USSAF in Europe, Office of the Deputy CG, May 30, 1945, in *History of the Army Air Forces Participation in Project Paperclip*, microfilm A2055, 919.

78. Memo to CG, USSAF in Europe, from HQ USSAF in Europe, deputy CG, June 1, 1945, in *History of the Army Air Forces Participation in Project Paperclip*, microfilm A2055, 920.

79. Memo to Major General Queede from D. L. Putt, Oct. 16, 1945, in *History of the Army Air Forces Participation in Project Paperclip*, microfilm A2055, 1008.

80. Memo to General Knerr from D. L. Putt, "Exploitation of German Scientists in the US," Nov. 28, 1945, in *History of the Army Air Forces Participation in Project Paperclip*, microfilm A2055, 1064–65.

81. Memo to CG, AAF, "Retention of PWs for Project Overcast," March 14, 1946, box 3693, RD 3723, RG 342, NARA.

82. Memo from CIC, USFET, Region V Landshut Sub-Region, signed Paul W. Witman, April 26, 1946, vol. 14, box 80, Records of the IIR, CIC, RG 319, NARA.

83. Memo to assistant chief of staff, G-2, First Infantry Division, from Norman T. Woods, First Lt. Inf., May 20, 1946, box 1, RG 338, NARA.

84. Memo to Col. Bixel, assistant chief of staff, Third Army, from Leonard Clark, Combined Arms Center, "Inspection of Operation Paperclip," April 13, 1946, box 1, RG 338, NARA.

85. Memo to director, JIOA, Monthly Report on Exploitation of German and Austrian Scientists and Technicians, Nov. 6, 1946, box 4, JIOA General Correspondence, RG 330, NARA.

86. Memo for director of JIOA, Aug. 19, 1946, box 1, JIOA General Correspondence, RG 330, NARA.

87. Gimbel, "German Scientists, US Denazification Policy," 446.

88. JIOA memo for director, JIOA, notes on conference with Lt. Col. Love, Civil Affairs Division, War Department Special Staff, concerning denazification of German nationals, July 31, 1947, box 2, JIOA General Correspondence, RG 330, NARA.

89. Quoted in Gimbel, "German Scientists, US Denazification Policy," 459.

90. Memo from War Department, Civil Affairs Division, to OMGUS, Aug. 7, 1947, box 19, Records of the IIR, CIC, RG 319, NARA.

91. Gerald Steinacher, *Nazis on the Run: How Hitler's Henchmen Fled Justice* (New York: Oxford University Press, 2011), xxiv.

92. Thomas Boghardt, "'Dirty Work?' The Use of Nazi Informants by US Army Intelligence in Postwar Europe," *Journal of Military History* 79 (April 2015): 394–96.

93. Memo to CO, 970th CI Det., HQ EUCOM, US Army, requests for security reports on, and investigations of, Paper Clip specialists, Sept. 5, 1947, box 19, Records of the IIR, CIC, RG 319, NARA.

94. Boghardt, "Dirty Work?," 399.

95. Ibid., 403.

96. Jacobsen, *Operation Paperclip*, 228.

97. Simpson, *Blowback*, 35.

98. JIOA, "Purpose of Security Reports for German Scientists Recommended for Immigration," to USFET G-2, Dec. 19, 1946, box 2, JIOA General Correspondence, RG 330, NARA.

99. Memo to chief, FIAT, from Science and Technology, "Report on Investigations Conducted to Determine Political Attitude of Richard Kuhn," Feb. 8, 1946, folder 2, 17-4, box 7, FIAT General Records, RG 260, NARA.

100. Memo to director, JIOA, December Report on Exploitation and Immigration of German and Austrian Scientists, Jan. 7, 1947, box 8, JIOA General Correspondence, RG 330, NARA.

101. Personal statement, Hans Hueter, May 16, 1947, Helmut Hueter, box 76, Foreign Scientist Case Files, RG 330, NARA.

102. Memo to chief, Administration and Liaison Group, Intelligence Division, JIOA, "Political Background of Paperclip Specialist Ernst Stuhlinger," June 6, 1949, Ernst Stuhlinger, box 164, Foreign Scientist Case Files, RG 330, NARA.

103. March 3, 1947 cable, box 2, JIOA General Correspondence, RG 330, NARA.

104. Ibid. See also memo for director, JIOA, February Report, Exploitation Division, March 4, 1947, box 8, JIOA General Correspondence, RG 330, NARA.

105. Memo to director, JIOA, from chief, Visa Division, Department of State, "Immigration of Paperclip Specialist Helmut Sieg," Dec. 13, 1949, box 24, JIOA Foreign Correspondence, RG 330, NARA.

106. Gerard J. Degroot, *Dark Side of the Moon: The Magnificent Madness of the American Lunar Quest* (New York: New York University Press, 2006), 25. JIOA dossiers with discrepancies include Ernst A. Steinhoff, Ernst Stuhlinger, Kurt Debus, and Hermann Kurzweg.

107. Mr. Cummings to Maj. Schmedmann, April 4, 1947, box 2, JIOA General Correspondence, RG 330, NARA.

108. Hunt, *Secret Agenda*, 115.

109. Oral history interview with Martin Mendelson, RG-50.702*0026, Hunt Collection, USHMM, accessed June 20, 2016, http://collections.ushmm.org/search/catalog/irn48109.

110. Tom Bower, "The Nazi Connection," *Frontline*, aired Feb. 24, 1987 (Chicago: FMI Films, 1987), VHS.

111. Oral history interview with Montee Cone, RG-50.702*0030, Hunt Collection, USHMM, accessed July 5, 2015, http://collections.ushmm.org/search/catalog/irn48113.

112. Revised security report on German (or Austrian) scientist or important technician from Office of the US High Commissioner for Germany, Jan. 8, 1950, Kurt Debus, box 28, Foreign Scientist Case Files, RG 330, NARA.

113. Memo to JIOA from EUCOM Intelligence, security report of Paperclip specialist Kurt Debus, Jan. 11, 1950, Kurt Debus, box 28, Foreign Scientist Case Files, RG 330, NARA.

114. Memo for chief, Intelligence Division, GSUSA, attn: chief, Special Projects Section, "Immigration of Paperclip Specialist Kurt Debus," June 27, 1949, box 28, Foreign Scientist Case Files, RG 330, NARA.

115. Memo to JIOA from EUCOM Intelligence, security report of Paperclip specialist Kurt Debus, Jan. 11, 1950, Kurt Debus, box 28, Foreign Scientist Case Files, RG 330, NARA.

116. Memo for the secretary of the army through assistant chief of staff, G-2, from Office of the Chief of Ordnance, "Immigration of Paperclip Specialist, Dr. Kurt Debus," June 9, 1950, Kurt Debus, box 28, Foreign Scientist Case Files, RG 330, NARA.

117. Hunt, *Secret Agenda*, 100.

118. Memo for chief, Collection Branch, Air Intelligence Requirements Division, director of intelligence, from JIOA, security report for Emil Salmon, Dec. 6, 1947, Emil Salmon, box 141, Foreign Scientist Case Files, RG 330, NARA.

119. Memo to director of intelligence, HQ USAF, from HQ AMC, Wright Patterson, German specialist Emil Salmon, April 28, 1950; and memo to chief, Collection Branch, Collection Division, director of intelligence, USAF, from JIOA, May 12, 1950, both in Emil Salmon, box 141, Foreign Scientist Case Files, RG 330, NARA.

120. Security report, Dec. 10, 1948, Emil Salmon, box 141, Foreign Scientist Case Files, RG 330, NARA.

121. Memo for the officer in charge from OMGUS, Land Württemberg Baden, Fort Military Government Battalion, US Army, alleged suspicion of subject's participation in the arson of the synagogue in Ludwigshafen, May 11, 1948, Emil Salmon, box 141, Foreign Scientist Case Files, RG 330, NARA.

122. Memo to EUCOM Special Projects Team from OMGUS, Land Württemberg Baden, 7780th OMGUS Group, Emil Salmon, June 15, 1949, Emil Salmon, box 141, Foreign Scientist Case Files, RG 330, NARA.

123. Security evaluation, April 19, 1949, Emil Salmon, box 141, Foreign Scientist Case Files, RG 330, NARA.

124. Memo to CG, AMC, from George D. Garrett Jr., director of intelligence, "Immigration of Paperclip Specialist, Ernst Eckert," Feb. 16, 1949, Ernst Eckert, box 34, Foreign Scientist Case Files, RG 330, NARA.

125. Memo to CG, USFET, Frankfurt, from JIOA, K. M. Schmedemann, Jan. 29, 1947, Ernst Eckert, box 34, Foreign Scientist Case Files, RG 330, NARA.

126. FBI file 77-2404, Ernst Rudolf Georg Eckert, June 9, 1946, 105-105284, Ernst Rudolf Georg Eckert, box 126, RG 65, NARA.

127. Sworn affidavit, Peter Pulz, April 7, 1949, Ernst Eckert, box 34, Foreign Scientist Case Files, RG 330, NARA.

128. Sworn affidavit, Ernst Eckert, Feb. 23, 1949, Ernst Eckert, box 34, Foreign Scientist Case Files, RG 330, NARA.

129. Memo to director of intelligence, WDGS, attn: JIOA, from Office of the Deputy Director of Intelligence, "Report on Allegations against, and Political Activities of, Ernst R. G. Eckert, Paperclip Specialist," June 9, 1947, Ernst Eckert, box 34, Foreign Scientist Case Files, RG 330, NARA.

130. Security report, Oct. 25, 1946, Ernst Eckert, box 34, Foreign Scientist Case Files, RG 330, NARA.

131. FBI file 77-2404, Ernst Rudolf Georg Eckert, June 9, 1946, 105-105284, Ernst Rudolf Georg Eckert, box 126, RG 65, NARA.

132. Takashi Nishiyama, "Cross-Disciplinary Technology Transfer in Trans-World War II Japan," *Comparative Technology Transfer and Society* 1, no. 3 (Dec. 2003): 307.

133. Summary report, "Atomic Bomb Mission, Investigation into Japanese Activity to Develop Atomic Power," HQ, First Technical Detachment, Sept. 30, 1945, box 1, Formerly Top Secret Nuclear Physics Correspondence File, 1947–1951, Scientific and Special Projects Group, Economic and Scientific Section, Supreme Command for the Allied Powers (SCAP), RG 331, Records of Allied Operational and Occupation Headquarters, World War II, NARA.

134. Nishiyama, "Cross-Disciplinary Technology Transfer," 317.

135. Memo from SSgt Bernie Bergevin, 309th Air Engineering Squadron, 35th Fighter Group, Johnson Army Air Base, to commanding officer, Wright Field Research Center, March 27, 1947, file no. 231.2, box 3151, RD no. 2810, Central Decimal Correspondence Files, 1919–1950, Sarah Clark Files, RG 342, NARA.

136. Memo, Takashi Shima (Japanese), interrogation of, June 18, 1947, file no. 231.2, box 3151, RD no. 2810, Central Decimal Correspondence Files, 1919–1950, Sarah Clark Files, RG 342, NARA.

137. Memo for the record, May 22, 1946, box 1, SCAP, RG 331, NARA.

138. Memo for the record, March 2, 1948, box 1, Scientific and Technical Division, Economic and Scientific Section, SCAP, RG 331, NARA.

139. "Travel of Japanese Scientists and Technicians," box 7249, Industrial Production and Construction Branch, Industry Division, Director of Production and Utilities, Economic and Scientific Section, SCAP, Topical File, 1945–1950, RG 331, NARA.

140. Sheldon H. Harris, *Factories of Death: Japanese Biological Warfare, 1932–1945, and the American Cover-Up* (New York: Routledge, 1994), 205–23.

141. Memo for file, May 3, 1949, box 1, SCAP, RG 331, NARA.

142. Nishiyama, "Cross-Disciplinary Technology Transfer," 318–20. See John Dower, *War Without Mercy: Race and Power in the Pacific War* (New York: Pantheon, 1987).

143. Col. Bryan Evans, assistant deputy director for R&D, US Army, to J. C. Green, Dec. 28, 1948, box 4, Office of Technical Services (OTS), RG 40, General Records of the Commerce Department, NARA.

144. Memo to General Marquat from H. C. Kelly, "Utilization of Japanese Scientists by the United States," Sept. 3, 1948, box 1, Base Labs in Japan; Usefulness of Japanese Scientists, SCAP, RG 331, NARA.

145. Memo to H. C. Kelly from I. Rabi, member, US Scientific Mission, "Limited Importation of Japanese Scientific Personnel to the US for Limited Periods of Two or Three Years," Dec. 8, 1948, box 1, SCAP, RG 331, NARA.

146. "W. R. Dornberger Dies, German Rocket Expert," *Washington Post*, July 2, 1980.

147. Memo to Capt. Baurbot, Control Branch, Munich, from P. M. Wilson, Enemy Personnel Exploitation Section (EPES), "Deployment of German Scientists and Technicians in the US and USA," June 3, 1946, folder 26, box 2, FIAT General Records, RG 260, NARA.

148. Memo by the reps of the British Chiefs of Staff, "Coordinated Exploitation of German Scientists and Technicians in the US and the UK," Sept. 24, 1945, folders 1 and 2, part 1, box 42, SWNCC 257, 12, RG 353, NARA.

149. Farquharson, "Governed or Exploited?," 30.

150. Ibid.

151. Farquharson, "Governed or Exploited?," 36. See Matthew Uttley, "Operation 'Surgeon' and Britain's Post-War Exploitation of Nazi German Aeronautics," *Intelligence and National Security* 17, no. 2 (Summer 2002): 1–26.

152. Memo to Baurbot from Wilson, June 3, 1946, RG 260, NARA.

153. British Chiefs of Staff memo, Sept. 24, 1945, RG 353, NARA.

154. Quoted in Ian Cobain, "How T-Force Abducted Germany's Best Brains for Britain," *Guardian*, Aug. 28, 2007.

155. Stewart Payne, "How Britain Put Nazis' Top Men to Work," *Telegraph*, Aug. 30, 2007.

156. "Nazi Scientists in Canada," *Constantine Report*, April 15, 2010, http://constantinereport.com/nazi-scientists-in-canada-w1947-cbc-radio-broadcast.

157. Ian Traynor, "UK Arranged Transfer of Nazi Scientists to Australia," *Guardian*, Aug. 16, 1999, https://www.theguardian.com/uk/1999/aug/17/iantraynor.

158. "Plan of Organization of Allied Control Authority in Germany," June 4, 1945, folder 5, 17-2, box 5, FIAT General Records, RG 260, NARA.

159. Letter to Sir Charles Darwin, Panel on Allocation of German Scientists and Technicians for Civilian Purposes in UK, Dec. 10, 1946, signed R. H. Ranger, Lt. Col. Sig. Corps Communications Section, box 4, JIOA General Correspondence, RG 330, NARA.

160. Memo to EPES FIAT (US) from EPES FIAT (UK), "Maintenance of Families of Germans Taken to UK for Employment," Aug. 7, 1946, folder 31-2, box 2, FIAT General Records, RG 260, NARA.

161. Memo to director, Office of Military Government, US Zone, from OMGUS Internal Affairs and Communications, Civil Administration Division, March 8, 1946, folder 31-2, box 2, FIAT General Records, RG 260, NARA.

162. Leffler, "American Conception," 368.

163. "Exploitation of German Scientists in Military Research," *Intelligence Review* 159 (Aug. 1949): 19.

164. Telegram, 867N.01/4-3046, acting secretary of state to the secretary of state, at Paris, June 20, 1946, in US State Department, *Foreign Relations of the United States, 1946*, vol. 5, *The British Commonwealth: Western and Central Europe* (Washington, DC: Government Printing Office, 1969), 570.

165. C. Office, assistant political adviser for Germany, USFET, to Robert Murphy, US political adviser for Germany, June 11, 1946, box 1, RG 84, Records of the Foreign Service Posts of the Department of State, NARA.

166. Memo to T Subdivision, G-2, SHAEF from HQ Sixth Army Group, Office of Assistant Chief of Staff, G-2, Activities of Securite Militaire, June 2, 1945, folder 8, 17-2, box 6, FIAT General Records, RG 260, NARA.

167. Memo, "German Scientists into Russian and French Territory," box 31, Recruitment, Records of the IIR, CIC, RG 319, NARA.

168. Lucius Clay, deputy military governor, to General Noiret, deputy commander-in-chief, French Group Control Council, July 27, 1946, box 1, RG 84, NARA.

169. Memo to Lt. Col. K. W. Jones, G-2, FIAT, from EPES, G-2, FIAT, Dr. Kamm, request for information, July 9, 1945, folder 18, 17-3, box 6, FIAT General Records, RG 260, NARA.

170. Memo to Integration and Planning Branch, FIAT, through Col. Osborne, from chief, FIAT, Dr. Kamm, request for information, July 13, 1945, folder 18, 17-3, box 6, FIAT General Records, RG 260, NARA.

171. EPES, FIAT, to Major L. C. Cross, RE, May 23, 1946, folder 31, box 2, FIAT General Records, RG 260, NARA.

172. The Turbo-Jet Engine Center in France, C-39-G, "Transfer of German Scientists, Technicians, Engineers, etc. . . . to France," box 29, Records Relating to War Crimes, Demilitarization, German Industries, and Scientific Research, ca. 1945–ca.1950, RG 466, Records of the US High Commissioner for Germany, NARA.

173. ETF 350 E7832, Military Intelligence Division, Great Britain, "Research in France by German Scientists," M. A., London, report no. R-4131-46, Oct. 11, 1946, box 127, Technical Files Relating to Foreign, Chemical, Radiological, and Biological Warfare Retired for Reference Purposes, ca. 1920–ca. 1986, RG 373, Records of the Defense Intelligence Agency, NARA.

Chapter 3 · Conscientious Objectors

1. Oral history interview with H. Graham Morison, by Jerry N. Hess, Aug. 4, 1972, transcript, Harry S. Truman Library and Museum, https://www.trumanlibrary.org/oralhist/morison.htm.

2. Memo to Colonel McCormack from Mr. Hutton, "Exploitation of German Specialists in Science and Technology in the US," box 6793, Central Decimal File, 862.542, RG 59, NARA.

3. James Forrestal to secretary of state James Byrnes, Jan. 9, 1945, box 6794, Central Decimal File, 862.542, RG 59, NARA.

4. W. L. Clayton, assistant secretary of state, to Jack McCloy, assistant secretary of war, June 21, 1945, box 6793, Central Decimal File, 862.542, RG 59, NARA.

5. Letter from Department of State Legal Adviser, Green H. Hackworth, no addressee, July 23, 1945, box 6793, Central Decimal File, 862.542, RG 59, NARA.

6. Lichtblau, *Nazis Next Door*, 2.

7. Memo for director of JIOA, "US Agencies' Requirements for Clearance of German and Austrian Specialists," May 21, 1946, box 2, JIOA General Correspondence, RG 330, NARA.

8. Department of State, Visa Division, comments on SWNCC 257/25, Nov. 7, 1946, box 6795, Central Decimal File, 862.542, RG 59, NARA. See also Lasby, *Project Paperclip*, 162–63.

9. Memo to McCormack from Hutton, "Exploitation of German Specialists in Science and Technology in the US," RG 59, NARA.

10. Hunt, *Secret Agenda*, 36.

11. Donald P. Steury, "The OSS and Project SAFEHAVEN," CIA Center for the Study of Intelligence, last updated June 27, 2008, https://www.cia.gov/library/center-for-the-study-of-intelligence/csi-publications/csi-studies/studies/summer00/art04.html.

12. Quoted in Martin Lorenz-Meyer, *Safehaven: The Allied Pursuit of Nazi Assets Abroad* (Columbia: University of Missouri Press, 2007), 32.

13. Ibid., 48. Klaus's rift with the FBI and J. Edgar Hoover personally began during Safehaven and continued into the 1950s.

14. "Did You Happen to See Samuel Klaus?" *Washington Times-Herald*, Dec. 14, 1944.

15. Obituary, Samuel Klaus, *New York Times*, Aug. 3, 1963.

16. Lorenz-Meyer, *Safehaven*, 168.

17. Ibid., 32.

18. Ibid., 48–49.

19. Confidential inquiry, US Civil Service Commission, Samuel Klaus, Nov. 4, 1944, Department of Treasury, Samuel Klaus, Official Personnel File, National Personnel Records Center.

20. Quoted in Lorenz-Meyer, *Safehaven*, 45.

21. Quoted ibid., 32.

22. Memo for Mr. Burke Knapp, US Group CC (Political Division), from HQ US Group Control Council, "Coordination of Investigations in Germany," June 14, 1945, folder 51, box 3, FIAT, RG 260, NARA.

23. State Department, "Requirements for Security in Immigration to the United States of German Scientists," box 6794, Central Decimal File, 862.542, RG 59, NARA.

24. Memo for Mr. Acheson from Sam Klaus, "Entry of 1,000 German Specialists and Their Families," entry DG, box 11, Interagency Working Group (IWG), RG 59, NARA.

25. Ibid.

26. Ibid.

27. "Efforts to Preserve German Power Outside Allied Occupation," Sept. 17, 1946, box 103, Samuel Klaus Files, RG 59, NARA.

28. State Department, "Requirements for Security in Immigration," RG 59, NARA.

29. Klaus memo, Sept. 25, 1946, box 103, Samuel Klaus Files, RG 59, NARA.

30. Ibid.

31. Oral history interview with Seymour Rubin, Jan. 6, 1997, RG-50.030*0449, USHMM, http://collections.ushmm.org/search/catalog/irn504934.

32. State Department, "Requirements for Security in Immigration," RG 59, NARA.

33. Klaus memo, Sept. 25, 1946, RG 59, NARA.

34. Gimbel, *Science, Technology, and Reparations*, 44–45.

35. Mr. Braden to Mr. Acheson, May 14, 1946, box 6794, Central Decimal File, 862.542, RG 59, NARA.

36. Ibid.

37. Memo to Mr. Acheson (draft), box 103, Samuel Klaus Files, RG 59, NARA.

38. Lasby, *Project Paperclip*, 160–61.

39. Spruille Braden, "The Germans in Argentina," *Atlantic Monthly*, April 1946, 39–41.

40. Quoted in Steinacher, *Nazis on the Run*, 219.

41. Quoted ibid., 221.

42. Memo to Colonel Booth from Sam Klaus, JIOA 1/14, box 6795, Central Decimal File, 862.542, RG 59, NARA.

43. Memo to Mr. Braden from Sam Klaus, Dec. 6, 1946, box 6795, Central Decimal File, 862.542, RG 59, NARA.

44. Memo to Mr. Acheson from Mr. Braden, JIOA 1/14, Jan. 6, 1947; and memo to General Hilldring from Mr. O'Sullivan (comments of Mr. Braden and Mr. Klaus on JIOA 1/14, Feb. 11, 1947), box 6795, Central Decimal File, 862.542, RG 59, NARA.

45. Acheson from Braden, JIOA 1/14, Jan. 6, 1947, RG 59, NARA.

46. State Department memo for the secretary, SWNCC, March 14, 1947, folder 2, part 2, box 43, SWNCC 257, RG 353, NARA.

47. Memo from State Department, Visa Division, March 24, 1948, box 6796, Central Decimal File, 862.542, RG 59, NARA.

48. Memo for secretary of war from Howard Petersen, July 21, 1946, box 18, 211 Scientists, RG 107, Record of the Office of the Secretary of War, NARA.

49. Memo by acting secretary of state to President Truman, Aug. 30, 1946, 862.542/9-346 in US State Department, *Foreign Relations of the United States, 1946*, 5:689.

50. George C. Marshall, secretary of state, to Kenneth C. Royall, secretary of war, Aug. 12, 1947, box 43, part 3, SWNCC 257, 33, RG 353, NARA.

51. Hunt, *Secret Agenda*, 111.

52. Tom Bower details the internecine bureaucratic warfare in a chapter entitled "The Irreconcilables: Diplomats and Soldiers," in *Paperclip Conspiracy*, 157–87.

53. Memo to Mr. Robinson from Herbert C. Cummings, "CON's Responsibility in the German Scientists Program," Feb. 25, 1947, box 6795, Central Decimal File, 862.542, RG 59, NARA.

54. Memo from Sam Klaus, JIOA meeting, April 24, 1946, entry DG, box 11, IWG, RG 59, NARA.

55. Ibid.

56. "State Department Problems: German Scientists," April 26, 1946, entry DG, box 11, IWG, RG 59, NARA.

57. Ibid.

58. Ibid.

59. Memo for the files: "German Scientists—Legal Problems," May 2, 1946, box 103, Samuel Klaus Files, RG 59, NARA.

60. Memo for the files, Sam Klaus, JIOA meeting, April 26, 1946, box 103, Samuel Klaus Files, RG 59, NARA.

61. Memo for the files, Sam Klaus, JIOA meeting, June 11, 1946, box 103, Samuel Klaus Files, RG 59, NARA.

62. Memo for the files, May 14, 1946, box 103, Samuel Klaus Files, RG 59, NARA.

63. Memo to Mr. Panuch from S. Klaus, German immigration, June 20, 1946, box 3, Sam Klaus Files, RG 59, NARA.

64. Memo for the files, Sam Klaus, JIOA meeting, June 25, 1946, box 3, Sam Klaus Files, RG 59, NARA.

65. Memo for the files, JIOA meeting, July 29, 1946, box 3, Sam Klaus Files, RG 59, NARA.

66. Memo to Mr. Acheson from Mr. Braden, JIOA 1/14, Feb. 3, 1947, box 6795, Central Decimal File, 862.542, RG 59, NARA.

67. Memo to Robinson from Cummings, "CON's Responsibility," Feb. 25, 1947, RG 59, NARA.

68. Memo from Hamilton Robinson, "The State Department Security Procedure for German Scientists' Cases," March 17, 1947, box 6795, Central Decimal File, 862.542, RG 59, NARA.

69. Memo from Samuel Klaus, HR 3663, the Gossett Bill, June 24, 1946, entry DG, box 11, IWG, RG 59, NARA.

70. Memo for Acheson from Klaus, "Entry of 1,000 German Specialists," RG 59, NARA.

71. Samuel Klaus, "German Scientists Program," July 17, 1946, entry DG, box 11, IWG, RG 59, NARA.

72. Memo from Sam Klaus, JIOA meeting, June 6, 1946, entry DG, box 11, IWG, RG 59, NARA.

73. Memo for Acheson from Klaus, "Entry of 1,000 German Specialists," RG 59, NARA.

74. Denial program, May 27, 1946, entry DG, box 11, IWG, RG 59, NARA.

75. Klaus, "German Scientists Program," July 17, 1947, RG 59, NARA.

76. Memo to Acheson (draft), RG 59, NARA.

77. Memo for the files, Sam Klaus, JIOA meeting, July 12, 1946, box 103, Samuel Klaus Files, RG 59, NARA.

78. Memo to secretary, JIOA, from State Department member, governing committee meeting of Feb. 27, 1947, memo dated March 5, 1947, box 103, Samuel Klaus Files, RG 59, NARA.

79. Memo for the files, JIOA, Feb. 27, 1947, box 103, Samuel Klaus Files, RG 59, NARA.

80. Klaus, "German Scientists Program," July 17, 1947, RG 59, NARA.

81. Ibid.

82. Memo to D. M. Ladd from A. H. Belmont, Senator Styles's charge that a State Department employee refused entrance into the US of one thousand German scientists, Aug. 17, 1950, section 11, box 59, 105-8090, RG 65, NARA.

83. Memo for director, JIOA, "Procedure for Clearance of German Specialists in the US," May 15, 1946, box 2, JIOA General Correspondence, RG 330, NARA.

84. JIOA memo for Captain G. A. Sinclair, Sept. 20, 1946, box 4, JIOA General Correspondence, RG 330, NARA.

85. Memo from Bosquet Wev, director, JIOA, to Hamilton Robinson, Office of Controls, State Department, March 17, 1948, box 2, JIOA General Correspondence, RG 330, NARA.

86. JIOA memo to CIC, EUCOM, director of intelligence, Dec. 4, 1947, box 7, JIOA General Correspondence, RG 330, NARA.

87. JIOA memo for Lt. General S. J. Chamberlain, director of intelligence, GSUSA, Delays encountered in the Department of Justice in processing the immigration into the US of German scientists, April 27, 1948, box 2, JIOA General Correspondence, RG 330, NARA.

88. Ibid.

89. Ibid.

90. Ibid.

91. Memo for Mr. Tamm, Mr. Ladd, Mr. Tolson, from FBI Dir. Hoover, May 11, 1948, section 4, 105-8090, box 49, RG 65, NARA.

92. Ordway and Sharpe, *Rocket Team*, 245.

93. Memo for Legislative and Liaison Division, WDGS, chief of Naval Operations, HR 6869, A Bill to Amend Immigration and Naturalization Laws, July 17, 1946 box 2, JIOA General Correspondence, RG 330, NARA.

94. Ibid.

95. Letter to Frank Fellows, chairman, Subcommittee on Immigration and Naturalization Service, House of Representatives, box 7, JIOA General Correspondence, RG 330, NARA.

96. Letter to Earl C. Michener, chairman, Committee on the Judiciary, House of Representatives, box 7, JIOA General Correspondence, RG 330, NARA.

97. Memo to director FBI from SAC Norfolk, Senator Styles Bridges' charges that a State Department employee refused entrance into the US of 1,000 scientists, Aug. 9, 1950, section 11, box 59, 105-8090, RG 65, NARA.

98. Statement of Captain B. N. Wev, US Navy, chairman, JIOA, June 27, 1947, section 11, box 59, 105-8090, RG 65, NARA.

99. Ibid.

100. Ibid.

101. Gallup Poll, box 18, Records of the OTS, RG 40, NARA.

102. Appendix: proposed press release of March 11, 1946, on Paperclip in Gimbel, *Science, Technology, and Reparations*, 188; "Nazis Sent to US as Technicians," *New York Times*, Jan. 4, 1947; and "Citizenship Opposed for Nazi Scientists," *New York Times*, Dec. 30, 1946.

103. "Nazis Sent to US as Technicians."

104. Press release draft, "Exploitation of Germany for Technological and Scientific Information," box 4, JIOA General Correspondence, RG 330, NARA.

105. Ibid.

106. Proposed press release to be issued by War Department as Joint State-War-Navy-Commerce-Justice Department release, box 7, JIOA General Correspondence, RG 330, NARA.

107. "Potential Exploitation of German Electronics," *Intelligence Review*, Jan. 30, 1947, 52.

108. "The Electric Gun: German Experiment with Electrically Launched Projectiles," *Intelligence Bulletin*, May 1946, 27; "The Nazi Kamikazes," *Intelligence Bulletin*, June 1946, 38; "Guided Missiles . . . The Weapon of the Future," *Intelligence Bulletin*, April 1946, 1–17.

109. "Guided Missiles," 2.

110. Major Robert A. Carr, "How German Experts Aid Our Research," *Army Information Digest* 4 (Oct. 1949): 18.

111. Ibid., 19.

112. Harry F. Byrd, "Hitler's Experts Work for US," *American Magazine* 145, no. 3 (1948): 38.

113. *History of Army Air Forces Participation in Project Paperclip* (May 1945–March 1947), Aug. 1948, microfilm A2055, 712–16.

114. Memo from Captain F. R. Duborg, USN, to director, JIOA, AAF report on Project Paperclip, Aug. 15, 1947, box 8, JIOA General Correspondence, RG 330, NARA.

115. Ibid.

116. Memo to Mr. Petersen from Dean Rusk, March 1, 1947, box 18, 211 Scientists, RG 107, NARA.

117. Handwritten note to Colonel McCarthy from Dean Rusk, box 18, 211 Scientists, RG 107, NARA.

118. *Iron Age*, Oct. 9, 1947, box 18, Records of the OTS, RG 40, NARA.

119. Article by Thomas F. Reynolds, March 21, 1947, box 18, Records of the OTS, RG 40, NARA; *Iron Age*, Oct. 9, 1947, RG 40, NARA.

120. Thomas F. Reynolds, "How Nazi Secrets Benefit USA," March 22, 1947, box 18, Records of the OTS, RG 40, NARA.

121. C. Montieth Gilpin, Society for the Prevention of World War III, to Henry Wallace, July 22, 1946, OTS, box 1, RG 40, NARA.

122. T. H. Tetens, *Know Your Enemy* (New York: Society for the Prevention of World War III, 1944).

123. Federation of American Scientists (FAS) to President Truman, Feb. 24, 1947, box 8, JIOA General Correspondence, RG 330, NARA.

124. Ibid.

125. "Our Scientists Say, 'Send Nazis Home,'" *Washington Daily News*, April 2, 1947.

126. Memo for Mr. Petersen from secretary of war Robert Patterson, March 24, 1947, box 18, 211 Scientists, RG 107, NARA. Patterson responded to NAACP chairman Walter White's protests personally, asking White to speak directly with him before going public and assuring White that Germans would receive no special treatment. See Secretary of War Robert Patterson to Walter White, Jan. 25, 1947, box 18, 211 Scientists, RG 107, NARA.

127. Report on FAS from War Department, March 11, 1947, box 18, 211 Scientists, RG 107, NARA.

128. Charles Wilber, Fordham University, to Harry Truman, March 25, 1947, box 6795, Central Decimal File, 862.542, RG 59, NARA.

129. Samuel A. Goudsmit, "German Scientists in Army Employment," *Bulletin of Atomic Scientists* 3, no. 2 (1947): 64.

130. Ibid., 67.

131. H. A. Bethe and H. S. Sack, "German Scientists in Army Employment: A Protest," *Bulletin of Atomic Scientists* 3, no. 2 (1947): 67.

132. "Use of German Scientists: Their Presence Here as Workers for Government Considered Unfortunate," *New York Times*, Sept. 15, 1947.

133. Morton M. Hunt, "The Nazis Who Live Next Door," *Nation*, July 23, 1949, 83.
134. Joachim Joesten, "This Brain for Hire," *Nation*, Jan. 11, 1947, 36.

Chapter 4 • Their Germans

1. Quoted in Lasby, *Project Paperclip*, 6.

2. In 1949, Venona project decryptions of Soviet intelligence reports revealed that Soviet agent Klaus Fuchs had penetrated the Los Alamos nuclear lab and provided Soviet intelligence incredibly detailed information on a plutonium bomb. Paul Maddrell notes that "the information was so good that it enabled an engineer to draw up a blueprint of the bomb; indeed, the first Soviet atomic bomb, tested on 29 August 1949, was an exact copy of the Alamogordo [New Mexico] weapon." Maddrell, *Spying on Science*, 28–29. See also Pavel V. Oleynikov, "German Scientists in the Soviet Atomic Project," *Nonproliferation Review* 7, no. 2 (2008): 1–30.

3. Oleynikov, "German Scientists," 1.

4. McDougall, *Heavens and the Earth*, 33–37.

5. Asif A. Siddiqi, "The Rocket's Red Glare: Technology, Conflict, and Terror in the Soviet Union," *Technology and Culture* 44, no. 3 (July 2003): 489; McDougall, *Heavens and the Earth*, 65.

6. Andreas Heinemann-Grüder, "Keinerlei Untergang," 45; McDougall, *Heavens and the Earth*, 44.

7. McDougall, *Heavens and the Earth*, 52.

8. Norman Naimark, *The Russians in Germany: A History of the Soviet Zone of Occupation, 1945–1949* (Cambridge: Belknap Press, 1995), 206.

9. Asif A. Siddiqi, "Russians in Germany: Founding the Post-War Missile Program," *Europe-Asia Studies* 56, no. 8 (Dec. 2004): 1134.

10. Ibid., 1135–36.

11. V. L. Sokolov, *Soviet Use of German Science and Technology, 1945–1946* (New York: Research Program on the USSR, 1955), 2.

12. Ibid., 6.

13. Report, n.d., box 31, Industry, Records of the IIR, CIC, RG 319, NARA.

14. Report of Russian Liaison FIAT on Liaison Mission with IG Farben Delegation (USSR) at Hoechst, Germany, Dec. 20, 1945, folder 8, 17-2, box 6, FIAT General Records, RG 260, NARA.

15. Memo to Brigadier T. J. Betts, deputy of AC of S, G-2, from CIOS, SHAEF, "Views of US Members on the Future of CIOS," May 29, 1945, folder 17, 17-3, box 6, FIAT General Records, RG 260, NARA.

16. Memo to director of intelligence from FIAT, "Agreement for the Interchange of Technical Information," n. d., folder 7, 17-19, box 34, FIAT General Records, RG 260, NARA.

17. Heinemann-Grüder, "Keinerlei Untergang," 45.

18. General of the Army Sokolovsky to Lt. General Clay, deputy military governor, Sept. 8, 1945, folder 40, box 2, FIAT General Records, RG 260, NARA.

19. Memo to USFET for FIAT Main for Integration and Planning Branch from OMGUS, signed Clay, Nov. 9, 1945, folder 12, 17-3, box 5, FIAT General Records, RG 260, NARA.

20. Division staff meeting minutes, July 27, 1946, *Minutes of the Division Staff Meetings of the US Group Control Council for Germany and the OMGUS, July 1944–August 1949*, microfilm, roll 1 (Washington, DC: University Publications of America, 1979).

21. McDougall, *Heavens and the Earth*, 45.

22. Dolores Augustine, "Wunderwaffen of a Different Kind: Nazi Scientists in East German Industrial Research," *German Studies Review* 29, no. 3 (Oct. 2006): 580.

23. Report: German professor of Zhukov's staff; Soviet mission to survey German inventions; arrest of industrialist, Jan. 11, 1946, box 31, Recruitment, Records of the IIR, CIC, RG 319, NARA.

24. Naimark, *Russians in Germany*, 209.

25. Translation of official notice of the administration of Weimar, box 2, JIOA General Correspondence, RG 330, NARA.

26. Report: German professor of Zhukov's staff, Jan. 11, 1946, RG 319, NARA.

27. Sokolov, *Soviet Use of German Science*, 26.

28. Report, n.d., EBF 974, 105-8090, RG 65, NARA.

29. Michael D. Gordon, *Red Cloud at Dawn: Truman, Stalin, and the End of the Atomic Monopoly* (New York: Farrar, Straus and Giroux, 2009), 128.

30. Asif A. Siddiqi, *Rocket's Red Glare: Spaceflight and the Russian Imagination, 1857–1957* (Cambridge: Cambridge University Press, 2010), 117.

31. Naimark, *Russians in Germany*, 220.

32. Sokolov, *Soviet Use of German Science*, 26–27.

33. Ibid., 30.

34. Quoted in Naimark, *Russians in Germany*, 221.

35. Returnee scientists debriefed by the CIA and FBI relate similar experiences on the night of October 21, 1946. See CIA report no. 314 on Dr. Wilhelm Keller and Rubenznoyne Chemical Factory, May 24, 1952, section 13, box 61, 105-8090, RG 65, NARA; CIA report no. 262, autobiographical report of Wilhelm Ernst Helias, returnee from the USSR, section 18, box 78, 105-8090, RG 65, NARA; and memo to FBI director from SAC, Los Angeles, Horst Waldemar Schneider, Feb. 22, 1972, section 1, box 226, 65-72324, RG 65, NARA.

36. Report from Federal Agency for the Protection of the Constitution, "Rocket Research, Employment of German V-2 Specialists in the Soviet Union," July 19, 1954, RG 319, CIC, Records of the IIR, box 28, vol. 3, folder 1, NARA.

37. Naimark, *Russians in Germany*, 220.

38. Siddiqi, "Russians in Germany," 127.

39. Maddrell, *Spying on Science*, 30–31.

40. Ibid., 31.

41. Quoted in Naimark, *Russians in Germany*, 219.

42. Heinemann-Grüder, "Keinerlei Untergang," 49.

43. Telegram, 740.00119 Control (Germany), US political adviser for Germany [Murphy] to the secretary of state, Nov. 3, 1946, in US State Department, *Foreign Relations of the United States, 1946*, 5:742.

44. Naimark, *Russians in Germany*, 225.

45. Memo, "German Scientists into Russian and French Territory," box 31, Records of the IIR, CIC, RG 319, NARA.

46. Naimark, *Russians in Germany*, 225.

47. Telegram, 740.00119 Control (Germany) / 10-3046, the US political adviser for Germany [Murphy] to the secretary of state, October 31, 1946, in US State Department, *Foreign Relations of the United States, 1946*, 5:740–41.

48. Quoted in Naimark, *Russians in Germany*, 226.

49. Telegram, 740.00119 Control (Germany) / 10-3146, the US political adviser for Germany [Murphy] to the secretary of state, Oct. 31, 1946, in US State Department, *Foreign Relations of the United States, 1946*, 5:742.

50. Quoted in Naimark, *Russians in Germany*, 228.

51. Bulkeley, *Sputnik Crisis*, 60.

52. Ibid., 64–65.

53. Siddiqi, "Russians in Germany," 1131.

54. Siddiqi, "Rocket's Red Glare," 487–88.

55. Quoted in McDougall, *Heavens and the Earth*, 45.

56. Quoted ibid., 46.

57. Ibid.

58. Agent report, Gröttrup Rocket Testing Institute, Moscow, July 16, 1948, folder 1, vol. 1, box 28, Records of the IIR, CIC, RG 319, NARA.

59. Gerhard Reisig interview, June 27, 1985, transcript Peenemünde Interviews Project, 1989–1990, NASM.

60. Ibid.

61. Agent report, Gröttrup Rocket Testing Institute, July 16, 1948, RG 319, NARA.

62. HQ 7707 EUCOM Intelligence Center, US Army, Technisches Büro no. 115, the former Institute Berlin, Sept. 1948, folder 1, vol. 1, box 28, Records of the IIR, CIC, RG 319, NARA.

63. "Rocket Research," July 19, 1954, RG 319, NARA.

64. McDougall, *Heavens and the Earth*, 53–54.

65. HQ EUCOM, Office of the Deputy Director of Intelligence, Junkers plant, Koethen, Feb. 27, 1948, folder 1, vol. 1, box 28, Records of the IIR, CIC, RG 319, NARA.

66. CIC Region I, "Recruiting of Former V-Waffen Technicians by the Russians," July 30, 1947, box 31, Recruitment, Records of the IIR, CIC, RG 319, NARA.

67. Strategic Services Unit, War Dept., Mission to Germany, USFET, Russian offers to German physicists, June 14, 1946, box 31, Recruitment, Records of the IIR, CIC, RG 319, NARA.

68. CIC Region III (Bad Neuheim), Operation Mesa, Nov. 2, 1946, box 31, Recruitment, Records of the IIR, CIC, RG 319, NARA. Letters smuggled out of the Soviet occupation zone and the Soviet Union were incorporated into military intelligence digests like the Army's *Intelligence Review*. See "Soviet Exploitation of German Scientists and Facilities," *Intelligence Review* no. 79 (Aug. 21, 1947): 66.

69. Memo from HQ CIC, USFET, Region III, situation report, Russian zone, Aug. 31, 1946, vol. 6, box 78, Records of the IIR, CIC, RG 319, NARA.

70. Memo from HQ CIC Region VIII, 970th CIC Detachment, "German Scientific Personnel Employed in the USSR, Economic Intelligence," July 15, 1947, vol. 10, box 78, Records of the IIR, CIC, RG 319, NARA.

71. Intelligence report, April 19, 1950, folder 2, vol. 3, Records of the IIR, CIC, RG 319, NARA.

72. One scientist "kidnapped" by the Soviets supposedly became "a Russophile, since he told the source that he was quite satisfied with the whole system." See EUCOM intelligence report, RG 319, folder 2, vol. 3, Records of the IIR, CIC, RG 319, NARA.

73. Memo, "V-Weapon Production in the Russian Zone of Germany," Nov. 5, 1946, folder 3, vol. 1, box 28, Records of the IIR, CIC, RG 319, NARA.

74. HQ Sub-Region Marburg, CIC Region III, Aktion AU, July 17, 1947, folder 2, vol. 1, box 28, Records of the IIR, CIC, RG 319, NARA.

75. HQ Sub-Region Kassel, CIC Region III, Lager Dora, Jan. 31, 1947, RG 319, NARA.

76. HQ USFET, CIC Detachment 970, Project 113/32 V-Bomb Manufacture, Nov. 6, 1946, folder 3, vol. 1, box 28, Records of the IIR, CIC, RG 319, NARA.

77. HQ Sub-Region Schwäbisch Hall, CIC Region I, concentration camp in the Russian zone, Jan. 30, 1947, folder 2, vol. 1, box 28, Records of the IIR, CIC, RG 319, NARA.

78. Daily report, Operation Mesa, box 31, Recruitment, Records of the IIR, CIC, RG 319, NARA.

79. HQ CIC, USFET Region III (Bad Nauheim), Operation Mesa, V-1, V-2, rocket scientists Kuehnert and Kaefer, Oct. 1, 1946, folder 3, vol. 1, box 28, Records of the IIR, CIC, RG 319, NARA.

80. USFET Region II, CIC, "Infiltration of German Scientists into the Russian Zone," Sept. 12, 1946, box 31, Recruitment, Records of the IIR, CIC, RG 319, NARA.

81. HQ CIC, European Theater, Region III (Bad Nauheim): Gröttrup Rocket Testing Institute in Moscow, Russia, April 27, 1948, RG 319, NARA.

82. Agent report, Gröttrup Rocket Testing Institute, July 16, 1948, RG 319, NARA.

83. Naimark, *Russians in Germany*, 228.

84. "Trackdown of the German Scientist," *New York Times*, Sept. 22, 1963.

85. Maddrell, *Spying on Science*, 207–8.

86. CIA report, Edith Stephanie Elizabeth Barwich, returnee from Agudzeri, July 20, 1955, section 27, box 82, 105-8090, RG 65, NARA.

87. "Rocket Research," July 19, 1954, RG 319, NARA.

88. Maddrell, *Spying on Science*, 208.

89. Siddiqi, "Russians in Germany," 142.

90. Augustine, "Wunderwaffen of a Different Kind," 580, 585.

91. Siddiqi, "Russians in Germany," 121.

92. Ibid., 133.

93. Ibid., 136.

94. Ibid., 121, 142.

95. Leffler, "American Conception," 359.

96. Quoted in Yergin, *Shattered Peace*, 138.

97. Telegram, George Kennan to George Marshall [Long Telegram], Feb. 22, 1946, George M. Elsey Papers, Harry S. Truman Administration File, Truman Library and Museum, https://www.trumanlibrary.org/whistlestop/study_collections/coldwar/documents/pdf/6-6.pdf.

98. Leffler, *Preponderance of Power*, 109.

99. Clark Clifford, *American Relations with the Soviet Union*, Sept. 24, 1946, Conway Files, Truman Papers, Truman Library and Museum, Independence, Missouri, https://www.trumanlibrary.org/4-1.pdf.

100. Leffler, "American Conception," 367.

101. Larry A. Valero, "The American Joint Intelligence Committee and Estimates of the Soviet Union, 1945–1947," CIA Center for the Study of Intelligence, last updated June 27, 2008, https://www.cia.gov/library/center-for-the-study-of-intelligence/csi-publications/csi-studies/studies/summer00/art06.html.

102. "ORE, 1, 23 July 1946, Soviet Foreign and Military Policy," in *Assessing the Soviet Threat: The Early Cold War Years*, ed. Woodrow J. Kuhns (McLean, VA: CIA Center for the Study of Intelligence, 1997), 56–66.

103. "ORE 3/1, 31 October 1946, Soviet Capabilities for the Development and Production of Certain Types of Weapons and Equipment," in Kuhns, *Assessing the Soviet Threat*, 87–88.

104. "The Soviets' Utilization of German Scientists," *Intelligence Review* no. 131 (Aug. 26, 1948): 58–61.

105. Lucius D. Clay, *Decision in Germany* (Garden City, NY: Doubleday, 1950), 354.

106. "ORE 22–48 Excerpt, 2 April 1948, Possibility of Direct Soviet Military Action During 1948," in Kuhns, *Assessing the Soviet Threat*, 187.

107. Yergin, *Shattered Peace*, 360.

108. Quoted ibid., 358.

109. Clay, *Decision in Germany*, 354–55.

110. Meeting notes, Nov. 18, 1948, box 13, JIOA General Correspondence, RG 330, NARA.

111. Memo to G-4 from G-3, Jan. 22, 1949, box 1, Formerly Top Secret Nuclear Physics Correspondence File, 1947–1951, Scientific and Special Projects Group, Economic and Scientific Section, SCAP, RG 331, NARA.

112. Memo from Economic and Scientific Section S to Civil Section, Base Laboratories in Japan, March 14, 1949, box 1, SCAP, RG 331, NARA.

113. HQ 970th CIC Detachment, USFET, Periodic Report of Theater-Directed CIC Operations, Jan. 2, 1947, vol. 6, box 78, Records of the IIR, CIC, RG 319, NARA.

114. Operation Mesa, box 31, Recruitment, Records of the IIR, CIC, RG 319, NARA.

115. Extract, CIC daily report, March 21, 1946, box 31, Recruitment, Records of the IIR, CIC, RG 319, NARA.

116. HQ 970th CIC Detachment, "Russian Recruitment of German Scientists," Sept. 2, 1947, box 31, Recruitment, Records of the IIR, CIC, RG 319, NARA.

117. Memo to CG, AAF, from D. L. Putt, Col. Air Corps, deputy CG, Intelligence, June 13, 1946, in *History of the Army Air Forces Participation in Project Paperclip*, microfilm A2055, 1255.

118. Memo to CG, AAF, "Acquisition of German Scientists for Exploitation in Project Paperclip," March 26, 1946, box 3693, RD 3723, RG 342, NARA.

119. Thomas Boghardt, "America's Secret Vanguard: US Army Intelligence Operations in Germany, 1944–47," *Studies in Intelligence* 57, no. 2 (June 2013): 10.

120. Upper Austria Sub-Detachment, 430th CIC Detachment, "Attempted Abduction of Russian Dr. in Steyr," Jan. 13, 1950, vol. 3, box 66, Records of the IIR, CIC, RG 319, NARA.

121. Heinz A. L. Hellmond to General Charles E. Saltzman, assistant secretary of state, Jan. 3, 1948, box 15, JIOA General Correspondence, RG 330, NARA.

122. CIC Region III, "Russian Propaganda in US and British Zones Directed at V-weapon Engineers," Dec. 27, 1946, box 31, Recruitment, Records of the IIR, CIC, RG 319, NARA.

123. Enclosure A, OMGUS FIAT memo to William S. Culbertson, box 2, JIOA General Correspondence, RG 330, NARA.

124. Memo to director, FBI, from SAC, El Paso, "Guided Missile Project, US Army," June 14, 1947, box 46, 105-8090, RG 65, NARA.

125. Memo to Honorable William L. Clayton from Colonel Gruhn, JIOA, "Importation of German Scientists and Technologists for Benefit of US Science and Industry," box 4, JIOA General Correspondence, RG 330, NARA.

126. Telegram from Buenos Aires (embassy) to secretary of state, July 30, 1948, box 6796, Central Decimal File, 862.542, RG 59, NARA.

127. Newspaper clipping, "German Scientists Find Reds More Co-operative than Allies," *Washington Daily News*, Dec. 9, 1948, box 94, 105-8090-A, Foreign Scientist Case Files, RG 330, NARA.

128. Newspaper clipping, "We Can't Afford to Send 500 Alien Scientists Home," *Washington News*, Feb. 9, 1950, box 94, 105-8090-A, Foreign Scientist Case Files, RG 330, NARA.

129. Memo, July 22, 1947, box 96, Miscellaneous, Samuel Klaus Papers, RG 59, NARA.

130. Memorandum for director of JIOA, Aug. 19, 1946, box 1, JIOA General Correspondence, RG 330, NARA.

131. Memo from G-2, May 11, 1948, box 70, Project Decimal File, 1946–1948, RG 319, NARA.

132. Memo to JCS, JIOA from EUCOM, "Security Standards for Paperclip Personnel," May 7, 1948, box 16, JIOA General Correspondence, RG 330, NARA.

133. Maddrell, *Spying on Science*, 277.

134. Richard J. Aldrich, *The Hidden Hand: Britain, America, and Cold War Secret Intelligence* (New York: Woodstock Press, 2002), 190–92.

135. CIA report, "Scientific Progress in the USSR," Oct. 30, 1953, section 19, box 77, 105-8090, RG 65, NARA.

136. CIA report, "Some Observations Concerning the Education of Technical Personnel in the Soviet Union," Dec. 19, 1955, section 29, box 84, 105-8090, RG 65, NARA.

137. Air intelligence information report, "Interrogation of Paperclip Specialist by Captain Richard O. Olney," Hubertus Strughold, Dec. 8, 1949, Hubertus Strughold, box 164, Foreign Scientist Case Files, RG 330, NARA.

138. CIA report, Jürgen Karl Heinrich Rottgardt, "German Scientists Returned from the USSR," Aug. 6, 1952, section 15, box 63, 105-8090, RG 65, NARA.

139. CIA report, autobiographical report of Ursula Schaefer, returnee from the USSR, Oct. 24, 1952, section 15, box 63, 105-8090, RG 65, NARA.

140. CIA report, autobiographical report of Kurt Thöm, German engineer returned from the USSR, Sept. 5, 1952, section 15, box 63, 105-8090, RG 65, NARA.

141. FBI file 77-963, Herbert Axster, by Richard R. Rogers, Oct. 6, 1949, box 45, 105-11452, RG 65, NARA.

142. Memo to director, FBI, from SAC Philadelphia, Georg Emil Knausenberger, Sept. 27, 1951, box 13, Georg Emil Knausenberger, 105-10683, RG 65, NARA.

143. CIA report, autobiographical report of Ernst Schaaf, German technician returned from the USSR, July 29, 1952, section 15, box 63, 105-8090, RG 65, NARA.

144. CIA report, autobiographical report of Eitel Fritz Spiegel, returnee from the USSR, April 10, 1953, section 17, box 77, 105-8090, RG 65, NARA.

145. Memo to Mr. A. H. Belmont from Mr. C. E. Hennrich, "German Scientists Interrogation Program," Intelligence Advisory Committee, Oct. 27, 1951, section 13, box 61, 105-8090, RG 65, NARA.

146. Memo to director, FBI, from SAC Baltimore, Hermann Herbert Kurzweg, Oct. 29, 1951, Hermann Herbert Kurzweg, 105-10524, RG 65, NARA.

147. Memo to director, FBI, from SAC Buffalo, Walter Robert Hugo Heinrich Dornberger, Dec. 29, 1951, box 30, Walter Robert Hugo Heinrich Dornberger, 105-11072, RG 65, NARA.

148. Memo to director, FBI, from SAC, Washington Field Office (WFO), Georg Hans Madelung, Aug. 30, 1951, box 3, Georg Hans Madelung, 105-10370, RG 65, NARA.

149. Henry Tolkmith, an expert in tabun and sarin gas and a Dow Chemical employee, was accused of passing information to the Soviet Union from Walter Schieber, the former chief of the Armaments Supply Office and an SS Brigadeführer. Schieber sought employment in the United States but settled for work in West Germany. See box 112, Henry Tolkmith, 105-11134, RG 65, NARA.

150. Quoted in Boghardt, "Dirty Work?," 397.

Chapter 5 · Paperclip Vindicated

1. Longden, *T-Force*, 344.

2. Quoted in Gimbel, *Science, Technology, and Reparations*, 5.

3. Ibid., 24–45.

4. George Meader, *Confidential Report to the Special Senate Committee Investigating the National Defense Program on the Preliminary Investigation of Military Government in the Occupied Areas of Europe, November 22, 1946* (Washington, DC, 1946), 25.

5. Memo to all district directors from Mackey, Dec. 8, 1952, RG 319, NARA.

6. Gimbel, *Science, Technology, and Reparations*, 95.

7. Memo to Colonel E. W. Gruhn from J. C. Green, Oct. 17, 1945, box 6, OTS, RG 40, NARA.

8. J. C. Green, Commerce, to William Clayton, assistant secretary of state for economic affairs, Jan. 21, 1946, box 6494, Central Decimal File, 862.542, RG 59, NARA.

9. Letter from secretary of war to secretary of state, n.d., box 4, JIOA General Correspondence, RG 330, NARA.

10. "Exploitation of Germany," RG 330, NARA.

11. Lasby, *Project Paperclip*, 231.

12. Memo for Captain Francis R. Duborg, Navy Technical Intelligence Center, from JIOA, "Repatriation of German Scientists to Germany," Nov. 27, 1946, box 4, JIOA General Correspondence, RG 330, NARA.

13. Memo to undersecretary of commerce from J. C. Green, "Continued Demand by Industry for German Scientists," Nov. 7, 1947, box 1, OTS, RG 40, NARA.

14. Memo to J. C. Green from Ray L. Hicks, two OTS functions to be mentioned in "Science and Public Policy," Sept. 24, 1947, box 1, OTS, RG 40, NARA.

15. Memo to secretary of state James Byrnes from Harriman, secretary of commerce, Jan. 23, 1947, box 6795, Central Decimal File, 862.542, RG 59, NARA.

16. JIOA memo for Robert Frye, OTS, Commerce, "Immigration of Dr. Werner Heisenberg," April 15, 1947, box 6, JIOA General Correspondence, RG 330, NARA.

17. *History of the Army Air Forces Participation in Operation Paperclip* (Dec. 1945–April 1948), vol. 2, microfilm A2055, 1587–88.

18. Memo for director, JIOA, from J. C. Green, Commerce, July 23, 1947, "Department of Commerce Cooperation with Armed Services under Operation Paperclip," box 6, JIOA General Correspondence, RG 330, NARA.

19. Ciesla, "Das 'Project Paperclip,'" 294.

20. Report ASW, "211 Scientists Represent Status of Operation Paperclip," April 16, 1947, box 18, 211 Scientists, RG 107, NARA.

21. J. S. Knowlson, Stewart-Warner Corporation, to Howard Petersen, March 27, 1947, box 18, 211 Scientists, RG 107, NARA.

22. Memo for the secretary of war, "German Scientists at Wright Field," Aug. 5, 1946, box 18, 211 Scientists, RG 107, NARA.

23. Memo to Mr. David Sommers, special assistant to secretary of war, from G. W. Lewis, GSC, box 18, 211 Scientists, RG 107, NARA.

24. JIOA memo for Captain Dennison, March 21, 1947, box 8, JIOA General Correspondence, RG 330, NARA.

25. Memo for Lt. Colonel Powers, "Replacement Implications of Paperclip Specialists in the Light of Increasing Salary Schedules," Oct. 20, 1949, box 18, JIOA General Correspondence, RG 330, NARA.

26. Memo to Ray L. Hicks, OTS, from Bosquet Wev, JIOA, "Value to the National Economy of the Paperclip Program," Feb. 9, 1949, box 1, OTS, RG 40, NARA.

27. Letter to JIOA from Philip Dontell, vice president Oklahoma Power and Propulsion Lab, June 17, 1949, box 22, JIOA General Correspondence, RG 330, NARA.

28. Carlton Murdock, dean of the university faculty, to Francis J. Brown, American Council on Education, Feb. 27, 1947, box 6795, Central Decimal File, 862.542, RG 59, NARA.

29. Assistant secretary Michael W. Straus, Department of Interior, to Colonel E. W. Gruhn, JIOA, Oct. 22, 1945, box 6, OTS, RG 40, NARA.

30. Memo from commander, US Naval Forces, Germany, to chief of naval operations, Oct. 7, 1948, box 15, JIOA General Correspondence, RG 330, NARA.

31. Report on German scientists from Anthony Horick, 700-F CIC Det., USAF, (provisional), March 5, 1948, box 12, Heinz Paul Beer, 105-10680, RG 65, NARA.

32. Curt Cardwell, *NSC 68 and the Political Economy of the Cold War* (Cambridge: Cambridge University Press, 2015).

33. JIOA memo for Geoffrey Lewis, Office of German Affairs, Department of State, request for information on responsibilities assumed by HICOG in connection with JCS directives on German scientists and technicians, Nov. 18, 1949, box 24, JIOA General Correspondence, RG 330, NARA.

34. Memo for record, "JIOA Coordination, Repatriation Program," Aug. 4, 1951, CIA Release Paperclip, RG 263, NARA. See Jacobsen, *Operation Paperclip*, 338–40.

35. Memo to all district directors from Mackey, Dec. 8, 1952, RG 319, NARA.

36. Memo to Mr. J. E. Codd, Project 63, box 29, Project Decimal File, 319.1 (May–Dec. 1953), RG 319, NARA.

37. Memo to assistant chief of staff, G-2, Army, Project 63, July 6, 1951, box 33, JIOA General Correspondence, RG 330, NARA.

38. Quoted in Hunt, "Nazis Who Live Next Door," 84.

39. Dir. JIOA to CSGID, "Professional Status of Former Paperclip Specialists Following Their Immigration," March 6, 1950, box 18, JIOA General Correspondence, RG 330, NARA.

40. JIOA memo for intelligence chiefs, "Procurement of Paperclip Specialists," April 6, 1949, box 23, JIOA General Correspondence, RG 330, NARA.

41. Eric Walker, executive secretary, to C. H. Nordstrom, SRD, HICOG, Feb. 20, 1951, box 33, JIOA General Correspondence, RG 330, NARA.

42. Memo draft, n.d., box 35, JIOA General Correspondence, RG 330, NARA.

43. JIOA to Carl Nordstrom, Chief Scientific Group, OMGUS, Feb. 4, 1949, box 22, JIOA General Correspondence, RG 330, NARA.

44. Nordstrom to Ellis, July 19, 1951, box 36, JIOA General Correspondence, RG 330, NARA.

45. "Suggested Procedure for a New 'Paperclip' Project," box 41, JIOA General Correspondence, RG 330, NARA.

46. AMC HQ to director, JIOA, "Contact between German Specialists," Sept. 15, 1950, box 25, JIOA General Correspondence, RG 330, NARA.

47. Biographical sketch, Ernst Stuhlinger, box 164, Foreign Scientists Case File, RG 330, NARA.

48. Memo for director, JIOA, from George D. Garrett Jr., USAF, "Unclassified Procurement of Paperclip Specialists," July 11, 1949, box 17, JIOA General Correspondence, RG 330, NARA. Dornberger's case is mentioned in Jacobsen, *Operation Paperclip*, 394.

49. Notes for meeting of JIOA with G-2, Feb. 1, 1955, box 32, Project Decimal File, 1954, RG 319, NARA.

50. Memo to SRD, Dr. C. H. Nordstrom, Special Projects Branch, "Some Dangers in the Present Situation of German Science and the Effects of American Policy," July 19, 1951, box 11, RG 466, NARA.

51. Ibid.

52. JIOA memo to Colonel W. R. Philip, Intelligence, EUCOM, Sept. 28, 1951, box 35, JIOA General Correspondence, RG 330, NARA.

53. Letter to Mr. W. Park Armstrong, Jr., special assistant, State Department Office of Intelligence, Aug. 3, 1951, CIA Release Paperclip, RG 263, NARA.

54. JIOA to Dr. Carl H. Nordstrom, SRD, Feb. 20, 1952, CIA Release Paperclip, RG 263, NARA.

55. Memo to [redacted] from [redacted], Project 63, March 18, 1952, CIA Release Paperclip, RG 263, NARA.

56. Conference held with Mr. Lewis in State Department, Colonel Ellis, Mr. Green, and Commander Welte, box 22, JIOA General Correspondence, RG 330, NARA.

57. Memo to Office of Science Adviser from Science Office, Bern, "US Program of Denial in Germany," Dec. 7, 1951, box 38, JIOA General Correspondence, RG 330, NARA.

58. Memo from J. B. Koepli, Project 63 of JIOA, box 11, DG, IWG, RG 59, NARA.

59. Quarterly report, Project 63, July 1 to Sept. 30, 1953, box 29, Project Decimal File, 319.1, Jan.–April 1953, RG 319, NARA.

60. Memo from J. B. Koepli, Project 63 of JIOA, RG 59, NARA.

61. State Department message from US policy adviser, Frankfurt, to secretary of state, June 23, 1951, for Nordstrom, box 36, JIOA General Correspondence, RG 330, NARA.

62. Translation of chapter 5 from *Forschung Heisst Arbeit und Bro*, Jan. 12, 1951, box 37, JIOA General Correspondence, RG 330, NARA.

63. Memo, July 22, 1948, box 96, Miscellaneous, Samuel Klaus Files, RG 59, NARA.

64. FBI memo for Mr. Clegg and Mr. E. A. Tamm from Director Hoover, Oct. 4, 1940, Samuel Klaus FBI File, Internet Archive, https://archive.org/details/foia_Klaus_Samuel-HQ-1.

65. Memo to Mr. Ladd from Mr. E. G. Fitch, Aug. 2, 1946, Samuel Klaus FBI File (FOIA), Internet Archive, https://archive.org/details/foia_Klaus_Samuel-HQ-1.

66. Mark Hove, *History of the Bureau of Diplomatic Security of the US Department of State* (Washington, DC: Department of State, 2011), 79–123.

67. Raymond J. Batvinis, *The Origins of FBI Counter-Intelligence* (Lawrence: University Press of Kansas, 2007), 55.

68. Tim Weiner, *Enemies: A History of the FBI* (New York: Random House, 2012), 144.

69. Memo from FBI director to SAC, Washington, Sept. 26, 1946, Samuel Klaus FBI File in FOIA: Klaus, Samuel-HQ-1, Internet Archive, https://archive.org/details/foia_Klaus_Samuel-HQ-1.

70. Memo to the FBI director from D. M. Ladd, Sept. 13, 1946, Samuel Klaus FBI File in FOIA: Klaus, Samuel-HQ-1, Internet Archive, https://archive.org/details/foia_Klaus_Samuel-HQ-1.

71. Memo, July 15, 1947, box 105, Samuel Klaus Files, RG 59, NARA.

72. Memo for the file, July 1, 1947, box 105, Samuel Klaus Files, RG 59, NARA.

73. Memo for the file, July 2, 1947, box 105, Samuel Klaus Files, RG 59, NARA.

74. Memo, July 11, 1947, box 105, Samuel Klaus Files, RG 59, NARA.

75. Memo, July 22, 1947, box 105, Samuel Klaus Files, RG 59, NARA.

76. Memo, July 9, 1947, box 105, Samuel Klaus Files, RG 59, NARA.

77. Memo, Feb. 11, 1947, box 96, Miscellaneous, Samuel Klaus Files, RG 59, NARA.

78. Lasby, *Project Paperclip*, 227–28.

79. Memo to Mr. Benton, assistant secretary of state, from Howland Sargeant, deputy to the assistant secretary, comments by Representative Busbey on the loyalty of State Department personnel, June 2, 1947, box 95, Busbey, Samuel Klaus Files, RG 59, NARA.

80. FOIA: Klaus, Samuel-HQ-1, enclosures to the bureau, re: Samuel Ezekial Klaus, WFO file no. 62-5309, Internet Archive, accessed May 13, 2016, https://archive.org/details/foia_Klaus_Samuel-HQ-1.

81. Lasby, *Project Paperclip*, 228.

82. Document fragment, immigration diary, n.d., box 2, JIOA General Correspondence, RG 330, NARA.

83. Analysis of extension of remarks by Congressman Busbey in *Congressional Record*, March 25, 1948, p. A, 1982 ff, box 105, Samuel Klaus Files, RG 59, NARA.

84. Klaus to Edgar M. Chenoweth, chairman, Subcommittee, House Committee on Expenditures, April 1, 1948, box 105, Samuel Klaus Files, RG 59, NARA.

85. Joseph McCarthy, speech at Wheeling, West Virginia, 1950, accessed May 13, 2016, http://wps.prenhall.com/wps/media/objects/108/110880/ch26_a5_d2.pdf.

86. Dean Acheson, *Present at the Creation: My Years in the State Department* (New York: Norton, 1969), 362, 369–70.

87. Memo to Mr. Bryan from CON___ G. H. Stewart Jr., March 17, 1950, box 96, Personnel, Samuel Klaus Files, RG 59, NARA.

88. Memo to Adrian S. Fisher from Samuel Klaus, "German Scientist Program Charges," Aug. 9, 1950, box 105, Samuel Klaus Files, RG 59, NARA.

89. *Congressional Record*, July 18, 1950 (copy), box 60, EBF 356, 105-8090, RG 65, NARA.

90. Memo to Fisher from Klaus, "German Scientist Program Charges," Aug. 9, 1950, RG 59, NARA.

91. Ibid.

92. Memo, no distribution, July 21, 1950, box 105, Samuel Klaus Files, RG 59, NARA.

93. Analysis of remarks by Senators Bridges and Ferguson in *Congressional Record*, July 18, 1950, p. 10651, box 105, Samuel Klaus Files, RG 59, NARA.

94. Memo, July 27, 1950, box 105, Samuel Klaus Files, RG 59, NARA.

95. Memo of conversation, June 1, 1951, box 105, Samuel Klaus Files, RG 59, NARA.

96. Bower, "The Nazi Connection."

97. Memo, no distribution, July 21, 1950, RG 59, NARA.

98. Rubin oral history, Jan. 6, 1997, USHMM.

99. Quoted in Bower, *Paperclip Conspiracy*, 162–63.

100. Rubin oral history, Jan. 6, 1997, USHMM.

101. Jacobsen, *Operation Paperclip*, 348–56.

102. Excerpt from the deputy director of intelligence's diary of May 8, 1952, CIA Release Paperclip, RG 263, NARA.

103. Dr. Edward L. Young, Board of Directors of the Physicians Forum, to the president, April 1, 1952, box 42, JIOA General Correspondence, RG 330, NARA.

104. Leslie, *Cold War and American Society*, 8.

105. McDougall, *Heavens and the Earth*, 105.

106. Neufeld, *Von Braun*, 219.

107. McDougall, *Heavens and the Earth*, 99.

108. Ernst Steinhoff to Colonel Holger N. Toftoy, Office Chief of Ordnance, R&D, Aug. 3, 1950, box 161, Ernst Steinhoff, Foreign Scientist Case Files, RG 330, NARA.

109. Letter from Major General E. L. Ford, Chief of Ordnance, no addressee, n.d., box 161, Ernst Steinhoff, Foreign Scientist Case Files, RG 330, NARA.

110. Oral history interview with Walter Häussermann, RG-50.702*0023, Hunt Collection, USHMM, accessed May 20, 2016, http://collections.ushmm.org/search/catalog/irn48106.

111. HQ Sub-Region Kassel, CIC Region III, "Scientific and Technical Cross References: Opposition Groups within Research Circles," April 30, 1947, vol. 5, box 77, Records of the IIR, CIC, RG 319, NARA.

112. Reisig interview, June 5–7, 1989, NASM.

113. Memo for chief, Rocket Branch, Ordnance R&D, Office Chief of Ordnance, from Major James P. Hamill, Ordnance, April 7, 1947, box 5, Herbert Axster, Foreign Scientist Case Files, RG 330, NARA.

114. FBI file 77-963, Herbert Felix Axster, by Richard R. Rogers, Oct. 4, 1949, box 45, 105-11452, RG 65, NARA.

115. Ellis, JIOA, to Peyton Ford, deputy attorney general, Herbert Axster, Jan. 19, 1950, box 29, JIOA General Correspondence, RG 330, NARA.

116. Intelligence information extracts, from J. Vondruska, TSNDA-2A, March 18, 1947, EBF 62, box 48, 105-8090, RG 65, NARA.

117. Memo to director of intelligence, WDGS, from CG, AAF, German specialist Hans Brede, July 25, 1947, EBF 62, box 48, 105-8090, RG 65, NARA.

118. FBI file Robert Heinrich Karl Paetz, Guenther Erwin Hintze, April 18, 1952, box 30, 105-11065, RG 65, NARA.

119. See Laney, *German Rocketeers*, for an excellent overview of the rocket team's integration into Huntsville.

120. FBI file 77-713, Hans Hueter, Feb. 17, 1949, box 28, Hans Herbert Hueter, 105-11020, RG 65, NARA.

121. Memo to Mr. A. H. Belmont from Mr. F. M. Baumgardner, German scientists having access to secret information, Redstone Arsenal, Huntsville, AL, Myron E. Huston, informant, May 15, 1953, section 17, box 77, 105-8090, RG 65, NARA.

122. FBI file 62-1649, Doctor Erich Traub, Feb. 8, 1950, box 12, Erich Traub, 100-102929, RG 65, NARA.

123. FBI file 77-684, Dieter Georg Ernst Grau, Jan. 17, 1949, Dieter Georg Grau, 105-10966, RG 65, NARA.

124. FBI file 62-1649, Doctor Erich Traub, Feb. 8, 1950, RG 65, NARA.

125. Report on Rudolph, Arthur; Louis, Hugo, from Martin S. Crow, 115th CIC Detachment, May 21, 1953, Rudolph, OSI, RG 60, NARA.

126. Memo to director, FBI, from SAC, Birmingham, Magnus von Braun, internal security, GE&R, July 9, 1951, box 6, Magnus von Braun, 62-81046, RG 65, NARA.

127. FBI file 77-724, Magnus von Braun, report by James P. O'Neil, Feb. 28, 1949, box 6, Magnus von Braun, 62-81046, RG 65, NARA.

128. Ordway et al., "A Memoir: From Peenemünde to USA: A Classic Case of Technology Transfer," *Acta Astronautica* 60 (2007): 36.

129. Quoted in Bulkeley, *Sputnik Crisis*, 74–82.

130. CIA report, "German Guided Missile Experts Potentially Available for Employment in the US," May 1, 1951, box 35, JIOA General Correspondence, RG 330, NARA.

131. Quoted in Ann Murkusen et al., *The Rise of the Gunbelt: The Military Remapping of Industrial America* (New York: Oxford University Press, 1991), 32.

132. Murkusen et al., *Rise of the Gunbelt*, 32.

133. Quoted in Jacobsen, *Operation Paperclip*, 262.

134. Neufeld, "Guided Missile and the Third Reich," 67.

135. Rees interview, Nov. 8, 1989, NASM.

136. Dieter J. Huzel, *Peenemünde to Canaveral* (Englewood Cliffs, NJ: Prentice Hall, 1962), 124.

137. Reisig interview, June 5–7, 1989, NASM.

138. Reisig interview, June 27, 1985, NASM.

139. "Reach for the Stars," *Time*, Feb. 17, 1958.

140. Quoted in Ordway and Sharpe, *Rocket Team*, 247.

141. Wang, *In Sputnik's Shadow*, 13.

142. Lasswell, *Essays on the Garrison State*, 59–60.

143. Mills, *Power Elite*, 220.

144. Michael J. Neufeld, "Wernher von Braun's Ultimate Weapon," *Bulletin of Atomic Scientists* 63, no. 4 (July 1, 2007): 52.

145. Speech before the Business Advisory Council for the Department of Commerce, Washington, DC, Sept. 17, 1952, in *The Voice of Dr. Wernher von Braun: An Anthology*, ed. Irene E. Powell (Burlington, Ontario: Apogee Books, 2007), 33.

146. Neufeld, "Wernher von Braun's Ultimate Weapon," 55–56.

147. David F. Noble, *The Religion of Technology: The Divinity of Man and the Spirit of Invention* (New York: Knopf, 1997), 124–28.

148. Wernher von Braun, "Why I Chose America," *American Magazine*, July 1952, 15, 111–12, 114–15.

149. Speech before the Huntsville Ministerial Association, St. Thomas Episcopal Church, Huntsville, Alabama, Nov. 13, 1962, in Powell, *Voice of Dr. Wernher von Braun*, 89.

150. Speech before the Associated General Contractors of America, Miami, Florida, January 1959, in Powell, *Voice of Dr. Wernher von Braun*, 36.

151. In *Inside Labor*, a column by Victor Riesel, Feb. 2, 1959 (copy), box 151, Wernher von Braun, 105-10747, RG 65, NARA.

152. Quoted in Launius, "Historical Dimension of Space Exploration," 25.

Epilogue

1. Robert A. Caro, *Master of the Senate: The Years of Lyndon Johnson III* (New York: Random House, 2003), 1024.

2. McDougall, *Heavens and the Earth*, 131.

3. Ibid., 157.

4. "East and West Wooed Germans," *New York Times*, July 17, 1969.

5. Comptroller General of the United States, *Nazis and Axis Collaborators Were Used to Further US Anti-Communist Objectives in Europe—Some Immigrated to the United States*

(Washington, DC: General Accounting Office, 1985), http://www.gao.gov/assets/150/142984.pdf.

6. Bower, "The Nazi Connection."

7. Kilgore, "Engineers' Dreams," 111.

8. Elford A. Cederberg, "German Scientists: Truman's Folly," *American Mercury* 87 (July 1958): 121.

9. "Reach for the Stars," *Time*, Feb. 17, 1958.

10. Quoted in Ordway and Sharpe, *Rocket Team*, 264.

11. Michael J. Neufeld, "The End of the Army Space Program: Interservice Rivalry and the Transfer of the von Braun Group to NASA, 1958–1959," *Journal of Military History* 69, no. 3 (July 2005): 742.

12. Ibid., 747.

13. T. Keith Glennan, *The Birth of NASA: The Diary of T. Keith Glennan* (Washington, DC: NASA History Office, 1993), 23.

14. Quoted in Ordway and Sharpe, *Rocket Team*, 274.

15. Mike Wright, "The Disney–Von Braun Collaboration and Its Influence on Space Exploration," in *Inner Space/Outer Space: Humanities, Technology, and the Postmodern World*, ed. Daniel Schenker, Craig Hanks, and Susan Kray (Huntsville, AL: Southern Humanities Press, 1993), 151–59.

16. De Witt Douglas Kilgore, *Astrofuturism: Science, Race, and Vision of Utopia in Space* (Philadelphia: University of Pennsylvania Press, 2003), 59–60.

17. Glennan, *Birth of NASA*, 6.

18. Michel, *Dora*, 98.

19. Jeff Stafford, "I Aim at the Stars," Turner Classic Movies Film Article, TCM, accessed June 22, 2016, http://www.tcm.com/this-month/article/246769%7C0/I-Aim-at-the-Stars.html.

20. Neufeld, "Creating a Memory," 73.

21. David H. DeVorkin and Michael J. Neufeld, "Space Artifact or Nazi Weapon? Displaying the Smithsonian's V-2 Missile, 1976–2011," *Endeavour* 35, no. 4 (2011): 192.

22. Michael J. Neufeld, "Smash the Myth of the Fascist Rocket Baron: East German Attacks on Wernher von Braun in the 1960s," in *Imagining Outer Space: European Astroculture in the Twentieth Century*, ed. Alexander C. T. Geppert (New York: Palgrave Macmillan, 2012), 107.

23. Paul Maddrell, "What We Have Discovered about the Cold War Is What We Already Knew: Julius Mader and the Western Secret Services during the Cold War," *Cold War History* 5, no. 2 (May 2005): 239.

24. Translation of advanced publication announcement, Aug. 16, 1963, Wernher von Braun, FBI file, FOIA, 432, https://vault.fbi.gov/Wernher%20VonBraun.

25. Lloyd W. Blankenbaker, NASA, director of security, to Hoover, June 12, 1964, Wernher von Braun, FBI file, FOIA, 430, https://vault.fbi.gov/Wernher%20VonBraun.

26. Neufeld, "Smash the Myth of the Fascist Rocket Baron," 116–17.

27. Quoted ibid., 118.

28. Neufeld, "Creating a Memory," 84.

29. Neufeld, *Von Braun*, 429.

30. Von Braun audio, RG-50.702*0010, Hunt Collection, USHMM.

31. Neufeld, "Wernher von Braun, the SS," 65.

32. Quoted in Laney, *German Rocketeers*, 149.

33. Feigin, *Office of Special Investigations*, 333.

34. Quoted in Laney, *German Rocketeers*, 153.

35. Oral history interview with Neal Sher, RG-50.702*0015, Hunt Collection, USHMM, accessed June 27, 2016, http://collections.ushmm.org/search/catalog/irn48086.

36. Feigin, *Office of Special Investigations*, 334.

37. Oral history interview with Herschal Auerbach, RG-50.702*0018, Hunt Collection, USHMM, accessed June 27, 2016, http://collections.ushmm.org/search/catalog/irn48090.

38. See the transcripts of all three interviews and discussions between the OSI and Arthur Rudolph's attorney in Thomas Franklin, *An American in Exile: The Story of Arthur Rudolph* (Huntsville, AL: Hugh McInnish, 1987).

39. Feigin, *Office of Special Investigations*, 334.

40. Ibid., 337.

41. See Häussermann interview, RG-50.702*0023; oral history interview with Konrad Dannenburg, Georg von Tiesenhausen, and Walter Häussermann, RG-50.702*0029, accessed June 27, 2016, http://collections.ushmm.org/search/catalog/irn48112; oral history interview with Georg von Tiesenhausen, RG-50.702*0022, accessed June 27, 2016, http://collections.ushmm.org/search/catalog/irn48105, all in Hunt Collection, USHMM.

42. Ordway, "Rudolph Case," B52.

43. Quoted in Laney, *German Rocketeers*, 171.

44. Ibid.

45. Oral history interview with Jane Mabry, RG-50.702*0008, Hunt Collection, USHMM, accessed June 27, 2016, http://collections.ushmm.org/search/catalog/irn48039.

46. Oral history interview with James Wall, RG-50.702*0009, Hunt Collection, USHMM, accessed June 27, 2016, http://collections.ushmm.org/search/catalog/irn48043.

47. Williams, *Tragedy of American Diplomacy*, 320.

48. Quoted in Neufeld, *Von Braun*, 473.

49. Quoted in Christopher Frayling, *Mad, Bad and Dangerous? The Scientists and the Cinema* (London: Reaktion Books, 2005), 98–99.

BIBLIOGRAPHY

Unpublished Primary Sources

Avalon Project. Lillian Goldman Law Library. Yale Law School, New Haven, Connecticut. http://avalon.law.yale.edu/subject_menus/nca_v4menu.asp.

Clifford, Clark. *American Relations with the Soviet Union*. Sept. 24, 1946. Conway Files. Truman Papers. Harry S. Truman Library and Museum, Independence, Missouri. https://www.trumanlibrary.org/4-1.pdf.

Elsey, George M., Papers. Harry S. Truman Administration File. Harry S. Truman Library and Museum, Independence, Missouri.

History of the Army Air Forces Participation in Project Paperclip. Microfilm A2055. Air Force Historical Research Agency, United States Air Force, Washington, DC.

Hunt, Linda. Collection. Oral History Interviews. United States Holocaust Memorial Museum (USHMM), Washington, DC. http://collections.ushmm.org/search/catalog/irn47923.

Jessel, Walter. "A Travelogue Through a Twentieth-Century Life: A Memoir." Unpublished manuscript, 1996. Private collection of Alfred Jessel.

Klaus, Samuel. FOIA: Klaus, Samuel-HQ-1. Enclosures to the Bureau, re: Samuel Ezekial Klaus. WFO file no. 62-5309. Internet Archive. https://archive.org/details/foia_Klaus_Samuel-HQ-1.

———. Official Personnel File. National Personnel Records Center, St. Louis, Missouri.

Messersmith, George S., Papers. University of Delaware, Newark. http://www.lib.udel.edu/ud/spec/findaids/html/mss0109.html.

Morison, H. Graham. Oral History Interviews. Harry S. Truman Library and Museum, Independence, Missouri. http://www.trumanlibrary.org/oralhist/morison.htm.

National Archives and Records Administration (NARA), College Park, Maryland

RG 40 General Records of the Commerce Department

RG 59 General Records of the Department of State

RG 60 General Records of the Department of Justice

RG 65 Records of the Federal Bureau of Investigation

RG 84 Records of the Foreign Service Posts of the Department of State

RG 107 Records of the Office of the Secretary of War

RG 165 Records of the War Department General and Special Staffs

RG 260 Records of US Occupation Headquarters, World War II

RG 263 Records of the Central Intelligence Agency

RG 319 Records of the Army Staff

RG 330 Records of the Office of the Department of Defense

RG 331 Records of Allied Operational and Occupation Headquarters, World War II

RG 342 Records of the United States Air Force Commands, Activities, and Organizations
RG 338 Records of US Army Operational, Tactical, and Support Organizations
RG 353 Interdepartmental and Intradepartmental Committees
RG 373 Records of the Defense Intelligence Agency
RG 466 Records of the US High Commissioner for Germany

National Security Archive. Gelman Library. George Washington University, Washington, DC. http://www2.gwu.edu/~nsarchiv.

Peenemünde Interviews Project, 1989–1990. Smithsonian National Air and Space Museum Archives (NASM), Washington, DC.

Rubin, Seymour. Oral history interview. Permanent Collection. United States Holocaust Memorial Museum (USHMM), Washington, DC. http://collections.ushmm.org/search/catalog/irn504934.

von Braun, Wernher. FBI file. FBI Records: The Vault. https://vault.fbi.gov/Wernher%20VonBraun.

Published Primary Sources

Acheson, Dean. *Present at the Creation: My Years in the State Department*. New York: Norton, 1969.

Béon, Yves. *Planet Dora: A Memoir of the Holocaust and the Birth of the Space Age*. Translated by Yves Béon and Richard L. Fague. Boulder, CO: Westview Press, 1997.

Bush, Vannevar. *Science: The Endless Frontier*. Washington, DC: Government Printing Office, 1945. https://www.nsf.gov/od/lpa/nsf50/vbush1945.htm.

Clay, Lucius D. *Decision in Germany*. Garden City, NY: Doubleday, 1950.

Dornberger, Walter. *V-2: The Nazi Rocket Weapon*. New York: Ballantine, 1954.

Glennan, T. Keith. *The Birth of NASA: The Diary of T. Keith Glennan*. Washington, DC: NASA History Office, 1993.

Goudsmit, Samuel A. *Alsos*. Los Angeles: Tomash, 1986.

Huzel, Dieter J. *Peenemünde to Canaveral*. Englewood Cliffs, NJ: Prentice Hall, 1962.

Kuhns, Woodrow J., ed. *Assessing the Soviet Threat: The Early Cold War Years*. McLean, VA: CIA Center for the Study of Intelligence, 1997.

Meader, George. *Confidential Report to the Special Senate Committee Investigating the National Defense Program on the Preliminary Investigation of Military Government in the Occupied Areas of Europe, November 22, 1946*. Washington, DC, 1946.

Michel, Jean. *Dora: The Nazi Concentration Camp Where Modern Space Technology Was Born and 30,000 Prisoners Died*. New York: Holt, Rinehart and Winston, 1979.

Minutes of the Division Staff Meetings of the US Group Control Council for Germany (USGCC) and the Office of Military Government for Germany (US) (OMGUS), July 1944–August 1949. Washington, DC: University Publications of America, 1979.

National Archives and Records Service. *United States Army Investigation and Trial Records of War Criminals,* USA vs. Kurt Andrae et al. *(and Related Cases) April 27, 1945–June 11, 1958*. Pamphlet describing M1079. Washington, DC: National Archives Microfilm Publications, 1981.

National Security Council, Executive Secretary. *United States Objectives and Programs for National Security: A Report to the National Security Council*. NSC 68. Washington, DC: National Security Council, April 12, 1950. https://www.trumanlibrary.org/whistlestop/study_collections/coldwar/documents/pdf/10-1.pdf.

Ordway, Frederick I., III, Werner K. Dahm, Konrad Dannenberg, Walter Häussermann, Gerhard Reisig, Ernst Stuhlinger, Georg von Tiesenhausen, and Irene Willhite. "A Memoir: From Peenemünde to USA: A Classic Case of Technology Transfer." *Acta Astronautica* 60 (2007): 24–47.

Powell, Irene E., ed. *The Voice of Dr. Wernher von Braun: An Anthology*. Burlington, Ontario: Apogee Books, 2007.
Smith, James Edward, ed. *The Papers of General Lucius Clay: Germany 1945–1949*. Vol. 1. Bloomington: Indiana University Press, 1974.
Tetens, T. H. *Know Your Enemy*. New York: Society for the Prevention of World War III, 1944.
US Army, Forces in the Branch European Theatre, Research and Intelligence Branch. *Peenemünde East, Through the Eyes of 500 Detained at Garmish*. Washington, DC: US Army, ca. 1945. https://archive.org/details/PeenemundeeasttooOMar.
US Foreign Economic Administration. *Report on the Ad Hoc Interdepartmental (War-Navy) Committee to Handle Foreign Economic Administration Projects on Project 4—The Post-Surrender Treatment of German Engineering and Research in the "Secret Weapon" Field*. Washington, DC: Foreign Economic Administration, July 10, 1945.
US State Department. *Foreign Relations of the United States, 1946*. Vol. 5, *The British Commonwealth: Western and Central Europe*. Washington, DC: Government Printing Office, 1969.
US War Department, Military Intelligence Division. *German Technical Aid to Japan: A Survey*. Washington, DC: War Dept., Aug. 31, 1945. https://archive.org/details/GermanTechnicalAidToJapan.
von Braun, Wernher. "Survey of Development of Liquid Rockets in Germany and Their Future Prospects." *Journal of the British Interplanetary Society* 10, no. 2 (March 1951): 75.
———. "Why I Chose America." *American Magazine*, July 1952, 15, 111–12, 114–15.
Wegener, Peter. P. *The Peenemünde Wind Tunnels: A Memoir*. New Haven, CT: Yale University Press, 1996.

Newspapers and Periodicals

American Magazine
American Mercury
Army Information Digest
Atlantic Monthly
Bulletin of Atomic Scientists
Daily Camera
Daily Mail
Foreign Affairs
Haaretz
The Guardian
Intelligence Bulletin
Intelligence Review
New York Times
The Nation
The Telegraph
Time
Washington Daily News
Washington Post
Washington Times-Herald

Secondary Sources

Aldrich, Richard J. *The Hidden Hand: Britain, America, and Cold War Secret Intelligence*. New York: Woodstock Press, 2002.
Allen, Charles R., Jr. "Hubertus Strughold, Nazi in USA: Atrocities in the Name of Medical Science." *Jewish Currents* 28 (Dec. 1974): 5–9.

Ash, Mitchell G. "Denazifying Scientists—and Science." In *Technology Transfer out of Germany after 1945*, edited by Matthias Judt and Burghard Ciesla, 61–80. New York: Routledge Press, 2013.

Augustine, Dolores. "Wunderwaffen of a Different Kind: Nazi Scientists in East German Industrial Research." *German Studies Review* 29, no. 3 (Oct. 2006): 579–88.

Bainbridge, William Sims. *The Spaceflight Revolution: A Sociological Study*. New York: Wiley, 1983.

Baldwin, Peter. *Hitler, the Holocaust and the Historians Dispute*. Boston: Beacon Press, 1990.

Bar-Zohar, Michael. *The Hunt for German Scientists*. Translated by Len Ortzen. London: Arthur Baker, 1967.

Batvinis, Raymond J. *The Origins of FBI Counter-Intelligence*. Lawrence: University Press of Kansas, 2007.

Beyerchen, Alan D. "German Scientists and Research Institutions in Allied Occupation Policy." *History of Education Quarterly* 22, no. 3 (Autumn 1982): 289–99.

———. *Scientists under Hitler: Politics and the Physics Community in the Third Reich*. New Haven, CT: Yale University Press, 1977.

———. "What We Now Know About Nazism and Science." *Social Research* 59, no. 3 (Fall 1992): 615–41.

Biddle, Wayne. *Dark Side of the Moon: Wernher von Braun, the Third Reich, and the Space Race*. New York: Norton, 2009.

Bijker, Wiebe E., Thomas P. Hughes, and Trevor Pinch. *The Social Construction of Technological Systems: New Directions in the Sociology and History of Technology*. Cambridge, MA: MIT Press, 2012.

Boghardt, Thomas. "America's Secret Vanguard: US Army Intelligence Operations in Germany, 1944–47." *Studies in Intelligence* 57, no. 2 (June 2013): 1–18.

———. " 'Dirty Work?' The Use of Nazi Informants by US Army Intelligence in Postwar Europe." *Journal of Military History* 79 (April 2015): 387–22.

Bower, Tom. "The Nazi Connection." *Frontline*. Aired Feb. 24, 1987. Chicago: FMI Films, 1987. VHS.

———. *The Paperclip Conspiracy: The Hunt for the Nazi Scientists*. Boston: Little, Brown, 1987.

Boyd, Carl, and Akihiko Yoshida. *The Japanese Submarine Force and World War II*. Annapolis, MD: Naval Institute Press, 2002.

Boyne, Walter J. "Project Paperclip." *Air Force Magazine* (June 2007): 70–74.

Bulkeley, Rip. *The Sputnik Crisis and Early United States Space Policy: A Critique of the Historiography of Space*. Bloomington: Indiana University Press, 1991.

Bury, Helen. *Eisenhower and the Cold War Arms Race: "Open Skies" and the Military-Industrial Complex*. London: I. B. Tauris, 2013.

Cardwell, Curt. *NSC 68 and the Political Economy of the Cold War*. Cambridge: Cambridge University Press, 2015.

Caro, Robert A. *Master of the Senate: The Years of Lyndon Johnson III*. New York: Random House, 2003.

Casey, Steven. "Selling NSC-68: The Truman Administration, Public Opinion, and the Politics of Mobilization, 1950–51." *Diplomatic History* 29, no. 4 (Sept. 2005): 655–90.

Ciesla, Burghard. "Das 'Project Paperclip'—deutsche Naturwissenschaftler und Techniker in den USA (1946 bis 1952)." In *Historische DDR-Forschung: Aufsätze und Studien*, edited by Jürgen Kocka, 287–301. Berlin: Akademie Verlag, 1993.

———. "German High Velocity Aerodynamics and Their Significance for the United States Air Force, 1945–1952." In *Technology Transfer out of Germany after 1945*, edited by Matthias Judt and Burghard Ciesla, 93–106. Amsterdam: Harwood Academic Publishers, 1996.

Collier, Basil. *The Battle of the V-Weapons, 1944–1945*. Yorkshire: Emfield Press, 1976.
Comptroller General of the United States. *Nazis and Axis Collaborators Were Used to Further US Anti-Communist Objectives in Europe—Some Immigrated to the United States*. Washington, DC: Government Accountability Office, June 28, 1985. http://www.gao.gov/assets/150/142984.pdf
Cornwell, John. *Hitler's Scientists: Science, War, and the Devil's Impact*. New York: Penguin, 2003.
Damms, Richard V. "James Killian, the Technological Capabilities Panel, and the Emergence of President Eisenhower's 'Scientific-Technological Elite.'" *Diplomatic History* 24, no. 1 (Winter 2000): 57–78.
Degroot, Gerard J. *Dark Side of the Moon: The Magnificent Madness of the American Lunar Quest*. New York: New York University Press, 2006.
DeVorkin, David H. *Science with a Vengeance: How the Military Created the US Space Sciences after World War II*. New York: Springer-Verlag, 1993.
———. "War Heads into Peace Heads: Holger N. Toftoy and the Public Image of the V-2 in the United States." *Journal of the British Interplanetary Society* 45 (1992): 439–44.
DeVorkin, David H., and Michael J. Neufeld. "Space Artifact or Nazi Weapon? Displaying the Smithsonian's V-2 Missile, 1976–2011." *Endeavour* 35, no. 4 (2011): 187–95.
Dick, Steven J. *Remembering the Space Age*. Washington, DC: NASA, 2008.
Dik, Alan Daso. "Operation LUSTY: The US Army Air Forces' Exploitation of the Luftwaffe's Secret Aeronautical Technology, 1944–45." *Aerospace Power Journal* 16 (Spring 2002): 28–40.
Dower, John. *War Without Mercy: Race and Power in the Pacific War*. New York: Pantheon, 1987.
Etheridge, Brian C. "*The Desert Fox*, Memory Diplomacy, and the German Question in Early Cold War America," *Diplomatic History* 32, no. 2 (April 2008): 207–38.
Etzold, Thomas H. "Organization for National Security, 1945–1950." In *Containment: Documents on American Foreign Policy and Strategy, 1945–1950*, edited by Thomas H. Etzold and John Lewis Gaddis, 8–18. New York: Columbia University Press, 1978.
Etzold, Thomas H., and John Lewis Gaddis, eds. *Containment: Documents on American Foreign Policy and Strategy, 1945–1950*. New York: Columbia University Press, 1978.
Farquharson, John. "Governed or Exploited? The British Acquisition of German Technology, 1945–48." *Journal of Contemporary History* 32, no. 1 (Jan. 1997): 23–42.
Feigin, Judy. *The Office of Special Investigations: Striving for Accountability in the Aftermath of the Holocaust*. Dec. 2008. Washington, DC: Department of Justice, 2010. https://www.justice.gov/sites/default/files/criminal/legacy/2011/03/14/12-2008osu-accountability.pdf.
Franklin, Thomas. *An American in Exile: The Story of Arthur Rudolph*. Huntsville, AL: Hugh McInnish, 1987.
Frayling, Christopher. *Mad, Bad and Dangerous? The Scientists and the Cinema*. London: Reaktion Books, 2005.
Geppert, Alexander C. T., ed. *Imagining Outer Space: European Astroculture in the Twentieth Century*. New York: Palgrave Macmillan, 2012.
Gerber, Amy, director. *My Grandfather Was a Nazi Scientist: Opa, Von Braun and Operation Paperclip*. Middleburg, VA: Flatcoat Films, 2010.
Geyer, Michael. "German Strategy in the Age of Machine Warfare, 1914–1945." In *Makers of Modern Strategy: From Machiavelli to the Nuclear Age*, edited by Peter Paret, 537–97. New York: Oxford University Press, 1986.
Gimbel, John. "The American Exploitation of German Technical Know-How after World War II." *Political Science Quarterly* 105, no. 2 (Summer 1990): 295–309.

———. *The American Occupation of Germany: Politics and the Military, 1945–1949*. Stanford, CA: Stanford University Press, 1968.

———. "German Scientists, US Denazification Policy, and the 'Paperclip Conspiracy.'" *International History Review* 12, no. 3 (Aug. 1990): 441–65.

———. "Project Paperclip: German Scientists, American Policy, and the Cold War." *Diplomatic History* 14, no. 3 (July 1990): 343–66.

———. *Science, Technology, and Reparations: Exploitation and Plunder in Postwar Germany*. Stanford, CA: Stanford University Press, 1990.

———. "US Policy and German Scientists: The Early Cold War Years." *Political Science Quarterly* 101, no. 3 (1986): 433–51.

Gordon, Michael D. *Red Cloud at Dawn: Truman, Stalin, and the End of the Atomic Monopoly*. New York: Farrar, Strauss and Giroux, 2009.

Grunden, Walter E., Yutaka Kawamura, Eduard Kolchinsky, Helmut Maier, and Masakatsu Yamazaki. "Laying the Foundation for Wartime Research: A Comparative Overview of Science Mobilization in National Socialist Germany, Japan, and the Soviet Union." *Osiris* 20 (2005): 79–106.

Haberer, Joseph. *Politics and the Community of Science*. New York: Van Nostrand Reinhold, 1969.

Harris, Sheldon H. *Factories of Death: Japanese Biological Warfare, 1932–1945, and the American Cover-Up*. New York: Routledge, 1994.

Heinemann-Grüder, Andreas. "'Keinerlei Untergang': German Armaments Engineers during the Second World War and in the Service of the Victorious Powers." In *Science, Technology and National Socialism*, edited by Monika Renneberg and Mark Walker, 30–50. Cambridge: Cambridge University Press, 1994.

Hooks, Gregory, and Gregory McLauchlan. "The Institutional Foundation of Warmaking: Three Eras of US Warmaking, 1939–1989." *Theory and Society* 21, no. 6 (Dec. 1992): 757–88.

Hove, Mark. *History of the Bureau of Diplomatic Security of the US Department of State*. Washington, DC: Department of State, 2011.

Hunt, Linda. "Project Paperclip, Designed to Give America's Military a Scientific Advantage, Was at one Crucial Time Run by a Spy." *Military History*, 14, no. 1 (April 1997).

———. *Secret Agenda: The US Government, Nazi Scientists, and Project Paperclip, 1945–1990*. New York: St. Martin's Press, 1991.

Irving, David. *The Mare's Nest*. London: William Kimber, 1964.

Jacobsen, Annie. *Operation Paperclip: The Secret Intelligence Program that Brought Nazi Scientists to America*. New York: Little, Brown, 2014.

Janiewski, Dolores E. "Eisenhower's Relationship with the 'Military-Industrial Complex.'" *Presidential Studies Quarterly* 41, no. 4 (Dec. 2011): 667–92.

Judt, Matthias, and Burghard Ciesla, eds. *Technology Transfer out of Germany after 1945*. Amsterdam: Harwood Academic Publishers, 1996; New York: Routledge Press, 2013.

Kershaw, Ian. *The Nazi Dictatorship: Problems and Perspectives of Interpretation*. London: Bloomsbury Academic, 2000.

Kilgore, De Witt Douglas. *Astrofuturism: Science, Race, and Visions of Utopia in Space*. Philadelphia: University of Pennsylvania Press, 2003.

———. "Engineers' Dreams: Wernher von Braun, Willy Ley, and Astrofuturism in the 1950s." *Canadian Review of American Studies* 27, no. 2 (1997): 103–31.

Kocka, Jürgen, ed. *Historische DDR-Forschung: Aufsätze und Studien*. Berlin: Akademie Verlag, 1993.

Laney, Monique. *German Rocketeers in the Heart of Dixie: Making Sense of the Nazi Past during the Civil Rights Era*. New Haven, CT: Yale University Press, 2015.

Lasby, Clarence G. *Project Paperclip: German Scientists and the Cold War*. New York: Atheneum, 1971.

Lasswell, Harold D. *Essays on the Garrison State*. New Brunswick, NJ: Transaction Publishers, 1997.

Launius, Roger D. "The Historical Dimension of Space Exploration: Reflections and Possibilities." *Space Policy* 16 (2000): 23–38.

Ledbetter, James. *Unwarranted Influence: Dwight D. Eisenhower and the Military-Industrial Complex*. New Haven, CT: Yale University Press, 2011.

Leffler, Melvyn P. "The American Conception of National Security and the Beginnings of the Cold War, 1945–48." *American Historical Review* 89, no. 2 (April 1984): 346–81.

———. *A Preponderance of Power: National Security, the Truman Administration, and the Cold War*. Stanford, CA: Stanford University Press, 1992.

Leslie, Stuart W. *The Cold War and American Science*. New York: Columbia University Press, 1993.

Lichtblau, Eric. *The Nazis Next Door: How America Became a Safe Haven for Hitler's Men*. New York: Houghton Mifflin, 2014.

Longden, Sean. *T-Force: The Forgotten Heroes of 1945*. London: Constable and Robinson, 2009.

Lorenz-Meyer, Martin. *Safehaven: The Allied Pursuit of Nazi Assets Abroad*. Columbia: University of Missouri Press, 2007.

MacDonald, Fraser. "Space and the Atom: On the Popular Geopolitics of Cold War Rocketry." *Geopolitics* 13 (2008): 611–34.

Maddrell, Paul. "British-American Scientific Intelligence Collaboration during the Occupation of Germany." *Intelligence and National Security* 15, no. 2 (2000): 74–94.

———. *Spying on Science: Western Intelligence in Divided Germany, 1945–1961*. New York: Oxford University Press, 2006.

———. "What We Have Discovered about the Cold War Is What We Already Knew: Julius Mader and the Western Secret Services during the Cold War." *Cold War History* 5, no. 2 (May 2005): 235–58.

McDonald, Michael, and Viorel Badescu, eds. *The International Handbook of Space Technology*. Berlin: Springer-Verlag, 2014.

McDougall, Walter A. *The Heavens and the Earth: A Political History of the Space Age*. New York: Basic Books, 1985.

McLauchlan, Gregory. "World War II and the Transformation of the US State: The Wartime Foundations of US Hegemony." *Sociological Inquiry* 67, no. 1 (Feb. 1997): 1–26.

Megargee, Geoffrey P., ed. *United States Holocaust Memorial Museum Encyclopedia of Camps and Ghettos, 1933–1945*. Bloomington: Indiana University Press, 2009.

Mills, C. Wright. *The Power Elite*. 1956. Reprint, Oxford: Oxford University Press, 2000.

Murkusen, Ann, Peter Hall, Scott Campbell, and Sabina Deitrick. *The Rise of the Gunbelt: The Military Remapping of Industrial America*. New York: Oxford University Press, 1991.

Naimark, Norman. *The Russians in Germany: A History of the Soviet Zone of Occupation, 1945–1949*. Cambridge: Belknap Press, 1995.

"Nazi Scientists in Canada." *Constantine Report*, April 15, 2010. http://constantinereport.com/nazi-scientists-in-canada-w1947-cbc-radio-broadcast.

Nazi War Crimes and Japanese Imperial Government Records Interagency Working Group (IWG). *Final Report to the United States Congress, April 2007*. College Park, MD: IWG, 2007. http://www.archives.gov/iwg/reports/final-report-2007.pdf.

Neufeld, Michael J. "Creating a Memory of the German Rocket Program for the Cold War." In *Remembering the Space Age*, edited by Steven J. Dick, 71–87. Washington, DC: NASA, 2008.

———. "The End of the Army Space Program: Interservice Rivalry and the Transfer of the von Braun Group to NASA, 1958–1959." *Journal of Military History* 69, no. 3 (July 2005): 737–57.

———. "The Guided Missile and the Third Reich: Peenemünde and the Forging of a Technological Revolution." In *Science, Technology and National Socialism*, edited by Monika Renneberg and Mark Walker, 51–71. Cambridge: Cambridge University Press, 1994.

———. "Hitler, the V-2, and the Battle for Priority, 1939–1943." *Journal of Military History* 57, no. 3 (July 1993): 511–38.

———. "Mittelbau Main Camp (aka Dora)." In *United States Holocaust Memorial Museum Encyclopedia of Camps and Ghettos, 1933–1945*, edited by Geoffrey P. Megargee, 966–71. Bloomington: Indiana University Press, 2009.

———. *The Rocket and the Reich: Peenemünde and the Coming of the Ballistic Missile Era.* Cambridge, MA: Harvard University Press, 1995.

———. "Rolf Engel vs. the German Army: A Nazi Career in Rocketry and Repression." *History and Technology* 13, no. 1 (1996): 53–72.

———. "Smash the Myth of the Fascist Rocket Baron: East German Attacks on Wernher von Braun in the 1960s." In *Imagining Outer Space: European Astroculture in the Twentieth Century*, edited by Alexander C. T. Geppert, 106–26. New York: Palgrave Macmillan, 2012.

———. *Von Braun: Dreamer of Space, Engineer of War*. New York: Vintage Books, 2007.

———. "Wernher von Braun, the SS, and Concentration Camp Labor: Questions of Moral, Political, and Criminal Responsibility." *German Studies Review* 25, no. 1 (2002): 57–78.

———. "Wernher von Braun's Pact with the Devil." *World War II* 22, no. 8 (Dec. 2007): 21–23.

Nishiyama, Takashi. "Cross-Disciplinary Technology Transfer in Trans-World War II Japan." *Comparative Technology Transfer and Society* 1, no. 3 (Dec. 2003): 305–27.

Noble, David F. *The Religion of Technology: The Divinity of Man and the Spirit of Invention.* New York: Knopf, 1997.

Oleynikov, Pavel V. "German Scientists in the Soviet Atomic Project." *Nonproliferation Review* 7, no. 2 (2008): 1–30.

Ordway, Frederick I., III. "Rudolph Case Should Be Reopened." *Aerospace America* 26 (Aug. 1988): B52.

Ordway, Frederick I., III, and Randy Liebermann, eds. *Blueprint for Space: Science Fiction to Science Fact.* Washington, DC: Smithsonian Institution Press, 1992.

Ordway, Frederick I., III, and Mitchell R. Sharpe. *The Rocket Team.* New York: Crowell, 1979; Burlington, Ontario: Apogee Books, 2003. Page numbers refer to the 2003 edition.

Oreskes, Naomi. "Science and the Origins of the Cold War." In *Science and Technology in the Global Cold War*, edited by Naomi Oreskes and John Krige, 11–30. Cambridge, MA: MIT Press, 2014.

Oreskes, Naomi, and John Krige, eds. *Science and Technology in the Global Cold War.* Cambridge, MA: MIT Press, 2014.

Paret, Peter, editor. *Makers of Modern Strategy: From Machiavelli to the Nuclear Age.* New York: Oxford University Press, 1986.

Petersen, Michael B. *Missiles for the Fatherland: Peenemünde, National Socialism, and the V-2 Missile.* Cambridge: Cambridge University Press, 2009.

Reitlinger, Gerald. *The SS: Alibi of a Nation, 1922–1945.* New York: Da Capo, 1989.

Renneberg, Monika, and Mark Walker, eds. *Science, Technology and National Socialism.* Cambridge: Cambridge University Press, 1994.

———. "Scientists, Engineers and National Socialism." In *Science, Technology and National Socialism*, edited by Monika Renneberg and Mark Walker, 1–29. Cambridge: Cambridge University Press, 1994.

Rossiter, Margaret E. "Science and Public Diplomacy since World War II." *Osiris* 1 (1985): 273–94.

Samuel, Wolfgang W. E. *American Raiders: The Race to Capture the Luftwaffe's Secrets.* Jackson: University Press of Mississippi, 2004.

Scalia, Joseph Mark. *Germany's Last Mission to Japan: The Failed Voyage of U-234*. Annapolis, MD: Naval Institute Press, 2000.

Schafft, Gretchen, and Gerhard Zeidler. *Commemorating Hell: The Public Memory of Mittelbau-Dora*. Urbana: University of Illinois Press, 2012.

Schenker, Daniel, Craig Hanks, and Susan Kray, eds. *Inner Space/Outer Space: Humanities, Technology, and the Postmodern World*. Huntsville, AL: Southern Humanities Press, 1993.

Scott, Phil. "Watson's Whizzers." *Aviation History*, July 2010, 36–41.

Sellier, Andrè. *A History of the Dora Camp*. Translated by Stephen Wright and Susan Taponier. Chicago: Ivan R. Dee, 2003.

Sherry, Michael. *Preparing for the Next War: American Plans for Postwar Defense, 1941–1945*. New Haven, CT: Yale University Press, 1977.

Siddiqi, Asif A. *Rocket's Red Glare: Spaceflight and the Russian Imagination, 1857–1957*. Cambridge: Cambridge University Press, 2010.

———. "The Rocket's Red Glare: Technology, Conflict, and Terror in the Soviet Union." *Technology and Culture* 44, no. 3 (July 2003): 47–51.

———. "Russians in Germany: Founding the Post-War Missile Program." *Europe-Asia Studies* 56, no. 8 (Dec. 2008): 1131–56.

Simpson, Christopher. *Blowback: The First Full Account of America's Recruitment of Nazis, and Its Disastrous Effect on our Domestic and Foreign Policy*. New York: Weidenfeld and Nicolson, 1988.

Sokolov, V. L. *Soviet Use of German Science and Technology, 1945–1946*. New York: Research Program on the USSR, 1955.

Steinacher, Gerald. *Nazis on the Run: How Hitler's Henchmen Fled Justice*. New York: Oxford University Press, 2011.

Steury, Donald P. "The OSS and Project SAFEHAVEN." CIA Center for the Study of Intelligence. Last updated June 27, 2008. https://www.cia.gov/library/center-for-the-study-of-intelligence/csi-publications/csi-studies/studies/summer00/art04.html.

Stolzfus, Nathan, and Henry Friedlander, eds. *Nazi Crimes and the Law*. Cambridge: Cambridge University Press, 2008.

Stuhlinger, Ernst, and Michael J. Neufeld. "Wernher von Braun and Concentration Camp Labor: An Exchange." *German Studies Review* 26, no. 1 (2003): 121–26.

Stuhlinger, Ernst, and Frederick I. Ordway III. *Wernher von Braun: Crusader for Space*. Malabar, FL: Krieger, 1994.

Szöllösi-Janze, Margit. "National Socialism and the Sciences: Reflections, Conclusions and Historical Perspectives." In *Science in the Third Reich*, edited by Margit Szöllösi-Janze, 1–36. New York: Berg, 2001.

———, ed. *Science in the Third Reich*. New York: Berg, 2001.

Telotte, J. P. "Disney in Science Fiction Land." *Journal of Popular Film and Television* 33, no. 1 (Spring 2005): 12–20.

Tooze, Adam. *The Wages of Destruction: The Making and Breaking of the Nazi Economy*. New York: Penguin, 2006.

"Transcript of US Army Interrogation of Arthur Rudolph in June 1947." Appendix A in Thomas Franklin, *An American Exile: The Story of Arthur Rudolph*, 172–74. Huntsville, AL: Hugh McInnish, 1987.

Uttley, Matthew. "Operation 'Surgeon' and Britain's Post-War Exploitation of Nazi German Aeronautics." *Intelligence and National Security* 17, no. 2 (Summer 2002): 1–26.

Valero, Larry A. "The American Joint Intelligence Committee and Estimates of the Soviet Union, 1945–1947." CIA Center for the Study of Intelligence. Last updated June 27, 2008. https://www.cia.gov/library/center-for-the-study-of-intelligence/csi-publications/csi-studies/studies/summer00/art06.html.

Walker, Mark. "The Nazification and Denazification of Physics." In *Technology Transfer out of Germany after 1945*, edited by Matthias Judt and Burghard Ciesla, 49–60. Amsterdam: Harwood Academic Publishers, 1996.

Wang, Zuoyue. *In Sputnik's Shadow: The President's Science Advisory Committee and Cold War America*. New Brunswick, NJ: Rutgers University Press, 2009.

Ward, Bob. *Dr. Space: The Life of Wernher von Braun*. Annapolis, MD: Naval Institute Press, 2009.

Weiner, Tim. *Enemies: A History of the FBI*. New York: Random House, 2012.

Welsome, Eileen. *The Plutonium Files: America's Secret Medical Experiments in the Cold War*. New York: Delta, 1999.

Wildt, Michael. *An Uncompromising Generation: The Nazi Leadership of the Reich Security Main Office*. Translated by Tom Lampert. Madison: University of Wisconsin Press, 2010.

Williams, William Appleman. *The Tragedy of American Diplomacy*. 1972. Reprint, New York: Norton, 2009. Page numbers refer to the 2009 edition.

Winter, Frank H. "Foundations of Modern Rocketry: 1920s and 1930s." In *Blueprint for Space: Science Fiction to Science Fact*, edited by Frederick I. Ordway III and Randy Liebermann, 95–105. Washington, DC: Smithsonian Institution Press, 1992.

Wolfe, Andrea J. *Competing with the Soviets: Science, Technology, and the State in Cold War America*. Baltimore, MD: Johns Hopkins University Press, 2013.

Wright, Mike. "The Disney-Von Braun Collaboration and Its Influence on Space Exploration." In *Inner Space/Outer Space: Humanities, Technology, and the Postmodern World*, edited by Daniel Schenker, Craig Hanks, and Susan Kray, 151–59. Huntsville, AL: Southern Humanities Press, 1993.

Yergin, Daniel. *Shattered Peace: The Origins of the Cold War and the National Security State*. Boston: Houghton Mifflin, 1977.

Zoglin, Richard. "Bob Hope's 10 Best Jokes." *Vulture*, Nov. 14, 2014. http://www.vulture.com/2014/11/10-best-bob-hope-jokes.html.

Page numbers in *italics* refer to photographs.

CPSIA information can be obtained
at www.ICGtesting.com
Printed in the USA
FSHW012259080920
73642FS